Validated Numerics

Validated Numerics

A Short Introduction to Rigorous Computations

Warwick Tucker

PRINCETON UNIVERSITY PRESS

PRINCETON AND OXFORD

Published by Princeton University Press, 41 William Street,
Princeton, New Jersey 08540
In the United Kingdom: Princeton University Press,
99 Banbury Road, Oxford OX2 6JX

First paperback printing, 2023
Paperback ISBN 9780691247656

The Library of Congress has cataloged the cloth edition as follows:

Tucker, Warwick, 1970–
Validated numerics: a short introduction to rigorous computations / Warwick Tucker.
  p. cm.
Includes bibliographical references and index.
ISBN 978-0-691-14781-9 (hardcover)
1. Numerical calculations—Verification.   2. Science—Data processing.   I. Title.
QA76.95.T83 2011
502′.85—dc22   2010040560

British Library Cataloging-in-Publication Data is available

This book has been composed in Times
press.princeton.edu

Typeset by S R Nova Pvt Ltd, Bangalore, India

It seems a pity, nevertheless, that mathematical rigour should have to be abandoned precisely at the point when a problem is reduced to arithmetic.

—RAMON E. MOORE, *Introduction to Algebraic Problems* (Topics in Interval Analysis, Oxford University Press, 1969)

Contents

Preface

Since the creation of the digital computer, numerical computations have played an increasingly fundamental role in modeling physical phenomena for science and engineering. With regards to computing speed and memory capacity, the early computers seem almost amusingly crude compared to their modern counterparts. Not surprisingly, today's students find it hard to believe that serious scientific work could be produced using such crude devices. True enough, the computations were rudimentary, and a significant portion of the total work took place before and after the computer did its share. Nevertheless, real-world problems were solved, and the speed-up due to the use of machines pushed the frontier of feasible computing tasks forward. Through a myriad of small developmental increments, we have now reached the production of Peta-flop/Peta-byte computers – an incredible feat which must have seemed completely unimaginable fifty years ago.

Despite this tremendous development in performance, very little has changed in the way computers actually *perform* their calculations. This state of affairs has led us to the rather awkward position where we can perform formidable computing tasks at very high speed, but where we do not have the capability to judge the validity of the final results. The question *"Are we just getting the wrong answers faster?"* is therefore a valid one, albeit slightly unkind to the great community of scientific computing, which has provided us with excellent numerical methods throughout the years.

Due to the inherent limitations of any finite-state machine, numerical computations are almost never carried out in a mathematically precise manner. As a consequence, they do not produce exact results, but rather approximate values that usually, but far from always, are near the true ones. In addition to this, external influences, such as an over-simplified mathematical model or a discrete approximation of the same, introduce additional inaccuracies into the calculations. As a result, even an seemingly simple numerical algorithm is virtually impossible to analyze with regards to its accuracy. To do so would involve taking into account every single floating point operation performed throughout the entire computation. It is somewhat amazing that a program *performing only two floating point operations* can be challenging to analyze! At speeds of one billion operations per second, any medium-sized program is clearly out of reach. This is a particularly valid point for complex systems, which require enormous models and very long computer runs. The grand example in this setting is weather prediction, although much simpler systems display the same kind of inaccessibility.

Fortunately, there are computational models in which approximate results are automatically provided with guaranteed error bounds. The simplest such model – *interval analysis* – was developed by Ramon Moore in the 1960's, see [Mo66]. At the time, however, computers were still at an early stage of development, and the additional costs associated with keeping track of the computational errors were deemed as too high. Furthermore, without special care in formulating the numerical algorithms, the produced error bounds would inevitably become overly pessimistic, and therefore quite useless.

Today, the development of interval methods has reached a high level of sophistication: tight error bounds can be produced – in many cases even faster than non-rigorous computations can provide an "approximation". As a testament to this, several highly non-trivial results in pure mathematics have recently been proved using computer-aided methods based on such interval techniques, see e.g. [Ha95], [Tu02], and [GM03]. We have now reached the stage where we can demand rigor *as well as* speed from our numerical computations. In light of this, it is clear that the future development of scientific computation must include techniques for performing *validated numerics*. It is also clear that mathematicians have been given a new, powerful research tool for challenging, non-linear problems.

This, in a nutshell, is the topic of the book before you.

Introduction

WHAT IS VALIDATED NUMERICS?

Validated numerics is the field aiming at bridging the gap between scientific computing and pure mathematics – between speed and reliability.

One strain of the field aims at pushing the frontiers of computer–aided proofs in mathematical analysis. This area of research deals with problems that cannot be solved by traditional mathematical methods. Typically, such problems have a global component (e.g. a large search space) as well as a non-linear ingredient (the output is not proportional to the input). These problems have traditionally been studied through numerical simulations, and therefore our knowledge of these lack the rigour demanded by a formal proof.

Validated numerics aims to bridge the gap between a numerically observed phenomenon, and its mathematical counter-part. This is achieved by developing a means of computing numerically yet with mathematical rigour. Validated numerics merges pure mathematics with scientific computing in a novel way: instead of computing *approximations* to sought quantities, the aim is to compute *enclosures* of the same. The width of such an enclosure gives a direct quality measurement of the computation, and can be used to adaptively improve the calculations at run-time. This fundamental change of focus results in efficient and extremely fault-tolerant numerical methods, ideal for the systematic study of complex systems. As such, validated numerics can play an instrumental role as it is the only way to certify that a numerical computation meets required error tolerances. As computer simulations are gradually replacing physical experiments, this type of certification is of utmost importance.

Many challenging problems share the same type of inaccessibility due to non-linearities affecting the global behaviour of the systems. Neither tools from pure mathematics nor scientific computation alone have been successful in establishing quantitative information for such complex systems. The field of validated numerics aims not only to develop the mathematical foundation needed to overcome these obstacles, but also to produce concrete numerical methods able to provide mathematical statements about such systems.

THE SCOPE AND AIM OF THIS BOOK

The main goal of this text is to introduce the reader to the field of validated numerics by providing a theoretical foundation supplemented with illuminating examples.

The target audience is undergraduate or graduate students new to the field, and who are not necessarily trained mathematicians. Some knowledge of programming is useful, but not strictly necessary. The restriction to the one-dimensional setting is a conscious one. It allows us to focus exclusively on simple, but yet interesting problems, without too much mathematical framework. The computer exercises are intended to entice the reader into actually discovering how simple it is to implement the methods, and to solve numerical problems with rigor.

FURTHER READING

There are several books that treat the topic of validated numerics, all with different scopes. Some of my favourites include the classic by Moore [Mo66], which has now been superseded by [Mo79, MK09], together with the implementation–geared [KM81] by Kulish and Miranker, and the comprehensive [AH83] by Alefeld and Herzberger. The two books by Neumaier [Ne90, Ne01] are a must, and the ones by Aberth [Ab88, Ab98] offer some interesting reading too. Complex–valued problems are considered in [PP98]. Some more recent good books are [WH03], by Walster and Hansen, aimed at global optimization (as is Kearfott's book [Ke96]), and [JK01], by Jaulin et. al., which focuses on constraint propagation. Of course, nowadays the World Wide Web has a lot to offer, and the interval community has its place there too, see http://www.cs.utep.edu/interval-comp/main.html for a vast collection of references, software, upcoming conferences, research groups, mailing lists, etc.

ACKNOWLEDGMENTS

I would like to thank all students who have helped me transform my somewhat erratic lecture notes into a more self-contained, polished manuscript. This has been a lengthy process, and without the enthusiasm of the participants of my course on validated numerics, I would not have come this far.

Warwick Tucker
Uppsala
April 27, 2010

Validated Numerics

Chapter One

Computer Arithmetic

IN THIS CHAPTER, we give an elementary overview of how (and what type of) numbers are represented, stored, and manipulated in a computer. This will provide insight as to why some computations produce grossly incorrect results. This topic is covered much more extensively in, for example, [Mu09], [Hi96], [Ov01], and [Wi63].

1.1 POSITIONAL SYSTEMS

Our everyday decimal number system is a *positional system* in base 10. Since computer arithmetic is often built on positional systems in other bases (e.g., 2 or 16),[1] we will begin this section by recalling how real numbers are represented in a positional system with an arbitrary integer base $\beta \geq 2$. Setting aside practical restrictions, such as the finite storage capabilities of a computer, any real number can be expressed as an infinite string

$$(-1)^\sigma (b_n b_{n-1} \cdots b_0.b_{-1}b_{-2} \cdots)_\beta, \tag{1.1}$$

where $b_n, b_{n-1}, \ldots$ are integers in the range $[0, \beta - 1]$, and $\sigma \in \{0, 1\}$ provides the sign of the number. The real number corresponding to (1.1) is

$$x = (-1)^\sigma \sum_{i=-\infty}^{n} b_i \beta^i$$
$$= (-1)^\sigma (b_n \beta^n + b_{n-1}\beta^{n-1} + \cdots + b_0 + b_{-1}\beta^{-1} + b_{-2}\beta^{-2} + \cdots).$$

If the number ends in an infinite number of consecutive zeros we omit them in the expression (1.1). Thus we write $(12.25)_{10}$ instead of $(12.25000\ldots)_{10}$. Also, we omit any zeros preceding the integer part $(-1)^\sigma (b_n b_{n-1} \ldots b_0)_\beta$. Thus we write $(12.25)_{10}$ instead of $(0012.25)_{10}$, and $(0.0025)_{10}$ instead of $(000.0025)_{10}$. Allowing for either leading or trailing extra zeros is called *padding* and is not common practice since it leads to redundancies in the representation.

Even without padding, the positional system is slightly flawed. No matter what base we choose, there are still real numbers that do not have a unique representation. For example, the decimal number $(12.2549999\ldots)_{10}$ is equal to $(12.255)_{10}$, and the binary number $(100.01101111\ldots)_2$ is equal to $(100.0111)_2$. This redundancy, however, can be avoided if we add the requirement that $0 \leq b_i \leq \beta - 2$ for infinitely many i.

Exercise 1.1. *Prove that any real number $x \neq 0$ has a unique representation (1.1) in a positional system (allowing no padding) with integer base $\beta \geq 2$ under the two conditions (a) $0 \leq b_i \leq \beta - 1$ for all i, and (b) $0 \leq b_i \leq \beta - 2$ for infinitely many i.*

Exercise 1.2. *What is the correct way to represent zero in a positional system allowing no padding?*

1.2 FLOATING POINT NUMBERS

When expressing a real number on the form (1.1), the placement of the decimal[2] point is crucial. The *floating point* number system provides a more convenient way to represent real numbers. A floating point number is a real number on the form

$$x = (-1)^{\sigma} m \times \beta^e, \tag{1.2}$$

where $(-1)^{\sigma}$ is the sign of x, m is called the *mantissa*,[3] and e is called the *exponent* of x. Writing numbers in floating point notation frees us from the burden of keeping track of the decimal point: it always follows the first digit of the mantissa. It is customary to write the mantissa as

$$m = (b_0.b_1 b_2 \ldots)_{\beta}$$

where, compared to the previous section, the indexing of the b_i has the opposite sign. As this is the standard notation, we will adopt this practice in what follows. Thus we may define the set of floating point numbers in base β as:

$$\mathbb{F}_{\beta} = \{(-1)^{\sigma} m \times \beta^e : m = (b_0.b_1 b_2 \ldots)_{\beta}\},$$

where, as before, we request that β is an integer no less than 2, and that $0 \leq b_i \leq \beta - 1$ for all i, and that $0 \leq b_i \leq \beta - 2$ for infinitely many i. The exponent e may be any integer.

Expressing real numbers in floating point form introduces a new type of redundancy. For example, the base 10 number 123 can be expressed as $(1.23)_{10} \times 10^2$, $(0.123)_{10} \times 10^3$, $(0.0123)_{10} \times 10^4$, and so on. In order to have unique representations for non-zero real numbers, we demand that the leading digit b_0 be non-zero, except for the special case $x = 0$. Floating point numbers satisfying this additional requirement are said to be *normal* or *normalized*.[4]

Exercise 1.3. *Show that a non-zero floating point number is normal if and only if its associated exponent e is chosen minimal.*

---

[2]Being picky, the expression "*base* point" or "*radix* point" is more appropriate, unless $\beta = 10$.

[3]The mantissa is sometimes referred to as the *significand* or, rather incorrectly, as the *fractional part* of the floating point number.

[4]This should not be confused with the number-theoretic notion of a normal number. There a number is normal to base β if every sequence of n consecutive digits in its β-expansion appears with limiting probability β^{-n}.

So far, we have simply toyed with different representations of the real numbers $\mathbb{R}$. As this set is uncountably infinite, whereas a machine can only store a finite amount of information, more drastic means are called for: we must introduce a new, much smaller set of numbers designed to fit into a computer that at the same time approximate the real numbers in some well-defined sense.

As a first step toward this goal, we restrict the number of digits representing the mantissa. This yields the set

$$\mathbb{F}_{\beta,p} = \{x \in \mathbb{F}_\beta : m = (b_0.b_1 b_2 \ldots b_{p-1})_\beta\}.$$

The number p is called the *precision* of the floating point system. It is a nice exercise to show that although $\mathbb{F}_{\beta,p}$ is a much smaller set than $\mathbb{F}_\beta$, it is countably infinite. This means that even $\mathbb{F}_{\beta,p}$ is too large for our needs. Note, however, that the restriction $0 \le b_i \le \beta - 2$ for infinitely many i becomes void in finite precision.

A *finite* set of floating point numbers can be formed by imposing a fixed precision, as well as bounds on the admissible exponents. Such a set is specified by four integers: the base β, the precision p, and the minimal and maximal exponents $\check{e}$ and $\hat{e}$, respectively. Given these quantities, we can define parameterized sets of *computer representable* floating point numbers:

$$\mathbb{F}_{\beta,p}^{\check{e},\hat{e}} = \{x \in \mathbb{F}_{\beta,p} : \check{e} \le e \le \hat{e}\}.$$

Exercise 1.4. *Show that $\mathbb{F}_{\beta,p}^{\check{e},\hat{e}}$ is finite, whereas $\mathbb{F}_{\beta,p}$ is countably infinite, and $\mathbb{F}_\beta$ is uncountably infinite, with*

$$\mathbb{F}_{\beta,p}^{\check{e},\hat{e}} \subset \mathbb{F}_{\beta,p} \subset \mathbb{F}_\beta.$$

(See appendix A for the different notions of infinite.)

Exercise 1.5. *How many normal numbers belong to the set $\mathbb{F}_{\beta,p}^{\check{e},\hat{e}}$?*

Using a base other than 10 forces us to have to rethink which numbers have a finite representation. This can cause some confusion to the novice programmer.

Example 1.2.1 *With $\beta = 2$ and $p < \infty$, the number $1/10$ is not exactly representable in $\mathbb{F}_{\beta,p}$. This can be seen by noting that*

$$\sum_{k=1}^{\infty} (2^{-4k} + 2^{-(4k+1)}) = \frac{3}{2}\left(\frac{1}{1-2^{-4}} - 1\right) = \frac{1}{10}.$$

Interpreting the first sum as a binary representation, we have

$$1/10 = (0.00011001100110011\ldots)_2 = (1.1001100110011\ldots)_2 \times 2^{-4}.$$

Since this is a non-terminating binary number, it has no exact representation in $\mathbb{F}_{2,p}$, regardless of the choice of precision p.

This example may come as a surprise to all who use the popular step-sizes 0.1 or 0.001 in, say, numerical integration methods. Most computers use $\beta = 2$ in their internal floating point representation, which means that on these computers $1000 \times 0.001 \ne 1$. This simple fact can have devastating consequences, for example, for very long integration procedures. More suitable step-sizes would be $2^{-3} = 0.125$ and $2^{-10} = 0.0009765625$, which are exactly representable in base 2. Example 1.4.1 further illustrates how sensitive numerical computations can be to these small conversion errors.

1.2.1 Subnormal Numbers

As mentioned earlier, without the normalizing restriction $b_0 \neq 0$, a non-zero real number may have several representations in the set $\mathbb{F}_{\beta,p}^{\check{e},\hat{e}}$. (Note, however, that most real numbers no longer have *any* representation in this finite set.) We already remarked that these redundancies may be avoided by normalization, that is, by demanding that all non-zero floating point numbers have a non-zero leading digit. To illustrate the concept of normalized numbers, we will study a small toy system of floating point numbers, illustrated in Figure 1.1.

Figure 1.1 The normal numbers of $\mathbb{F}_{2,3}^{-1,2}$.

It is clear that the floating point numbers are not uniformly distributed: the positive numbers are given by

$$\{0.5, 0.625, 0.75, 0.875, 1, 1.25, 1.5, 1.75, 2, 2.5, 3, 3.5, 4, 5, 6, 7\}.$$

Thus the intermediate distances between consecutive numbers range within the set $\{0.125, 0.25, 0.5, 1\}$. In Section 1.3, we will explain why this type of non-uniform spacing is a good idea. We will also describe how to deal with real numbers having modulus greater than the *largest positive normal number* N_{max}^n, which in our toy system is $(1.11)_2 \times 2^2 = 7$.

Note that the *smallest positive normal number* in $\mathbb{F}_{2,3}^{-1,2}$ is $N_{min}^n = (1.00)_2 \times 2^{-1} = 0.5$, which leaves an undesired gap centered around the origin. This leads not only to a huge loss of accuracy when approximating numbers of small magnitude but also to the violation of some of our most valued mathematical laws (see Exercise 1.18). A way to work around these problems is to allow for some numbers that are not normal.

A non-zero floating point number in $\mathbb{F}_{\beta,p}^{\check{e},\hat{e}}$ with $b_0 = 0$ and $e = \check{e}$ is said to be *subnormal* (or denormalized). Subnormal numbers allow for a *gradual underflow* to zero (compare Figures 1.1 and 1.2). Extending the set of admissible floating point numbers to include the subnormal numbers still preserves uniqueness of representation, although the use of these additional numbers comes with some penalties, as we shall shortly see. For our toy system at hand, the positive subnormal numbers are $\{0.125, 0.25, 0.375\}$.

Figure 1.2 illustrates the normal and subnormal numbers of $\mathbb{F}_{2,3}^{-1,2}$. The difference between these two sets is striking: the *smallest positive normal number* N_{min}^n is $(1.00)_2 \times 2^{-1} = 0.5$, whereas the *smallest positive subnormal number* N_{min}^s is

$(0.01)_2 \times 2^{-1} = 0.125$. The largest subnormal number N^s_{max} is $(0.11)_2 \times 2^{-1}$ $= 0.375$. Geometrically, introducing subnormal numbers corresponds to filling the gap centered around the origin with evenly spaced numbers. The spacing should be the same as that between the two smallest positive normal numbers.

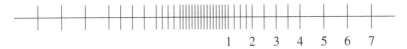

Figure 1.2 The normal and subnormal numbers of $\mathbb{F}^{-1,2}_{2,3}$.

From now on, when we refer to the set $\mathbb{F}^{\check{e},\hat{e}}_{\beta,p}$, we mean the set of normal *and* subnormal numbers. We will use $\mathbb{F}$ to denote any set of type $\mathbb{F}^{\check{e},\hat{e}}_{\beta,p}$ or $\mathbb{F}_{\beta,p}$; when needed, the exact parameters of the set in question will be provided. A real number x with $N^n_{min} \leq |x| \leq N^n_{max}$ is said to be in the *normalized range* of the associated floating point system.

Exercise 1.6. *Prove that the non-zero normal and subnormal numbers of $\mathbb{F}^{\check{e},\hat{e}}_{\beta,p}$ have unique representations.*

Exercise 1.7. *How many positive subnormal numbers are there in the set $\mathbb{F}^{\check{e},\hat{e}}_{\beta,p}$?*

Exercise 1.8. *Derive formulas for N^s_{min}, N^s_{max}, N^n_{min}, and N^n_{max} for a general set of floating point numbers $\mathbb{F}^{\check{e},\hat{e}}_{\beta,p}$.*

We conclude this section by remarking that although the floating point numbers are not uniformly spaced; for $\check{e} \leq m, n < \hat{e}$, the sets $\mathbb{F}^{\check{e},\hat{e}}_{\beta,p} \cap [\beta^m, \beta^{m+1})$ and $\mathbb{F}^{\check{e},\hat{e}}_{\beta,p} \cap [\beta^n, \beta^{n+1})$ have the same cardinality. This is apparent in Figures 1.1 and 1.2. We also note that any floating point system $\mathbb{F}$ is symmetric with respect to the origin: $x \in \mathbb{F} \Leftrightarrow -x \in \mathbb{F}$.

Exercise 1.9. *Compute the number of elements of the set $\mathbb{F}^{\check{e},\hat{e}}_{\beta,p} \cap [\beta^n, \beta^{n+1})$, provided that $\check{e} \leq n < \hat{e}$.*

1.3 ROUNDING

We have now reached the stage where we have succeeded in condensing the uncountable set of real numbers $\mathbb{R}$ into a finite set of floating point numbers $\mathbb{F}$. Almost all commercial computers use a set like $\mathbb{F}$, with some minor additions, to approximate the real numbers. In order to make computing feasible over $\mathbb{F}$, we must find a means to associate any real number $x \in \mathbb{R}$ to a member y of $\mathbb{F}$. Such an association is called *rounding* and defines a map from $\mathbb{R}$ onto $\mathbb{F}$. Obviously, we cannot make the map invertible, but we would like to make it as close as possible to a homeomorphism.

Before defining such a mapping, we will extend both the domain and range into the sets $\mathbb{R}^* = \mathbb{R} \cup \{-\infty, +\infty\}$ and $\mathbb{F}^* = \mathbb{F} \cup \{-\infty, +\infty\}$, respectively. This provides an elegant means for representing real numbers that are too large in absolute value to fit into $\mathbb{F}$. In the actual implementation, the symbols $\{-\infty, +\infty\}$ are specially coded and do not have a valid representation such as (1.2).

Following the excellent treatise on computer arithmetic, [KM81], a rounding operation $\bigcirc: \mathbb{R}^* \to \mathbb{F}^*$ should have the following properties:

(R1) $x \in \mathbb{F}^* \Rightarrow \bigcirc(x) = x$,
(R2) $x, y \in \mathbb{R}^*$ and $x \le y \Rightarrow \bigcirc(x) \le \bigcirc(y)$.

Property (R1) simply states that all members of $\mathbb{F}^*$ are fixed under $\bigcirc$. Clearly an already representable number does not require any further rounding. Property (R2) states that the rounding is monotone. Indeed, it would be very difficult to interpret the meaning of any numerical procedure without this property. Combining (R1) and (R2), one can easily prove that the rounding $\bigcirc$ is of *maximum quality*, that is, the interior of the interval spanned by x and $\bigcirc(x)$ contains no points of $\mathbb{F}^*$.

LEMMA 1.10. *Let* $x \in \mathbb{R}^*$. *If* $\bigcirc: \mathbb{R}^* \to \mathbb{F}^*$ *satisfies both (R1) and (R2), then the interval spanned by x and $\bigcirc(x)$ contains no points of $\mathbb{F}^*$ in its interior.*

Proof. The claim is trivially true if $x = \bigcirc(x)$ (since then the interior is empty), so assume that $x \ne \bigcirc(x)$. Without loss of generality we may assume that $x < \bigcirc(x)$. Now suppose that the claim is false, that is, there exists an element $y \in \mathbb{F}^*$ with $x < y < \bigcirc(x)$. Since, by (R1), we have $\bigcirc(y) = y$, we must, by (R2), have $\bigcirc(x) \le y$. This gives the desired contradiction. $\square$

We will now describe four rounding modes that are available on most commercial computers.

1.3.1 Round to Zero

The simplest rounding operation to implement is *round toward zero*, often referred to as *truncation*, which we denote by $\square_z$. We formally define $\square_z: \mathbb{R}^* \to \mathbb{F}^*$ by

$$\square_z(x) = \text{sign}(x) \max\{y \in \mathbb{F}^*: y \le |x|\}, \tag{1.3}$$

where $\text{sign}(x)$ is the sign of x. The action of $\square_z$ is illustrated in Figure 1.3. To see how easy this rounding mode is to implement, consider a real number $x = (-1)^\sigma (b_0.b_1 b_2 \ldots)_\beta \times \beta^e$ in $\mathbb{F}_\beta$ to be rounded into $\mathbb{F}_{\beta,p}$. If x satisfies $|x| \le N_{max}^n$, this is achieved by simply discarding the mantissa digits beyond position $p - 1$ (hence the nickname truncation): $\square_z(x) = (-1)^\sigma (b_0.b_1 b_2 \ldots b_{p-1})_\beta \times \beta^e$. Otherwise, x is rounded to $(-1)^\sigma N_{max}^n$.

Round toward zero is an *odd*[5] function:

(R3) $x \in \mathbb{R}^* \Rightarrow \bigcirc(-x) = - \bigcirc(x)$.

Rule (R3) is also satisfied by the most common rounding mode: *round to nearest*, which we shall return to later.

---

[5] A function f is said to be *odd* if $f(-x) = -f(x)$ and *even* if $f(-x) = f(x)$. Most functions are neither.

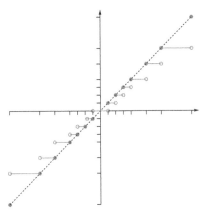

Figure 1.3 Round to zero $\square_z$.

1.3.2 Directed Rounding

There are two very useful rounding modes that are said to be *directed*. By this we mean that they satisfy (in addition to (R1) and (R2)) one of the following rules:

(R4) (a) $x \in \mathbb{R}^* \Rightarrow \bigcirc(x) \leq x$ or   (b) $x \in \mathbb{R}^* \Rightarrow \bigcirc(x) \geq x$.

The rounding mode satisfying (R4a) is called *round toward minus infinity* (or simply *round down*). The rounding mode satisfying (R4b) is called *round toward plus infinity* (or simply *round up*). These rounding modes are denoted $\nabla(x)$ and $\triangle(x)$, respectively, and are formally defined by

$$\nabla(x) = \max\{y \in \mathbb{F}^*: y \leq x\} \quad \text{and} \quad \triangle(x) = \min\{y \in \mathbb{F}^*: y \geq x\}. \quad (1.4)$$

The number $\nabla(x)$, called x *rounded down*, is the largest floating point number less than or equal to x, whereas $\triangle(x)$, called x *rounded up*, is the smallest floating point number greater than or equal to x. Note that if $x \in \mathbb{F}^*$, then $\nabla(x) = x = \triangle(x)$, whereas if $x \notin \mathbb{F}^*$, we have the enclosure $\nabla(x) < x < \triangle(x)$, which is of maximal quality. This means that the interval $[\nabla(x), \triangle(x)]$ contains no points of $\mathbb{F}^*$ in its interior (apply Lemma 1.10 twice). Also note the anti-symmetry relations:

$$\triangle(-x) = -\nabla(x) \quad \text{and} \quad \nabla(-x) = -\triangle(x). \quad (1.5)$$

Thus either rounding ∇ or $\triangle$ can be completely defined in terms of the other.

Exercise 1.11. *Using only (1.4), prove that the rounding modes ∇ and $\triangle$ satisfy (R1) and (R2). Also show that ∇ satisfies (R4a) and that $\triangle$ satisfies (R4b).*

Exercise 1.12. *Show that, in terms of the directed rounding mode ∇, we have*

$$\square_z(x) = \text{sign}(x) \nabla(|x|).$$

The relations (1.4) completely define the directed roundings ∇ and $\triangle$ as maps from $\mathbb{R}^*$ to $\mathbb{F}^*$ (see Figure 1.4).

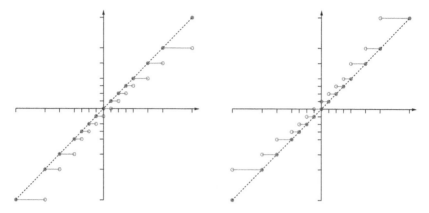

Figure 1.4 Directed roundings: (left) round down ∇; (right) round up $\triangle$.

1.3.3 Round to Nearest (Even)

Note that all previously defined rounding modes map the interior of any interval spanned by two consecutive floating point numbers onto a single point in $\mathbb{F}^*$. This means that the error made when rounding a single real number x could be as large as the length of the interval $[\nabla(x), \triangle(x)]$ enclosing it. A more accurate family of rounding modes is called *round to nearest*.

For an element of $x \in \mathbb{R}^*$, we can construct an enclosure of x in $\mathbb{F}^*$:

$$\nabla(x) \leq x \leq \triangle(x).$$

If also $|x| \leq N_{max}^n$, we let $\mu = \frac{1}{2}(\triangle(x) + \nabla(x))$ denote the midpoint of this interval. If $|x| > N_{max}^n$, we let $\mu = \text{sign}(x) N_{max}^n$. Rounding to nearest returns $\nabla(x)$ if $x < \mu$, and $\triangle(x)$ if $x > \mu$. In the rare case $x = \mu$, there is a tie. The different variants of rounding to nearest are distinguished according to how they resolve this tie.

One easy way to break the tie is to simply round up for positive ties and down for negative ones. This rounding mode is called *round to nearest* and is defined by

$$x > 0 \Rightarrow \square_n(x) = \begin{cases} \nabla(x), & \text{if } x \in [\nabla(x), \mu), \\ \triangle(x), & \text{if } x \in [\mu, \triangle(x)], \end{cases} \tag{1.6}$$

$$x < 0 \Rightarrow \square_n(x) = -\square_n(-x).$$

Although easy to describe, this rounding mode has the slight disadvantage of being *biased*: the rounding errors are not evenly distributed around zero. Indeed, if x is positive, then $\square_n$ has a higher probability of rounding x downward and vice versa (see Figure 1.5a).

An unbiased way to break the tie gives rise to a rounding mode called *round to nearest even*, which we simply denote by $\square$. This is the default rounding mode on almost all commercial computers. In order to define this rounding operation, we will assume that the mantissas of $\nabla(x)$ and $\triangle(x)$ are given by

$$(a_0.a_1 a_2 \ldots a_{p-1})_\beta \quad \text{and} \quad (b_0.b_1 b_2 \ldots b_{p-1})_\beta,$$

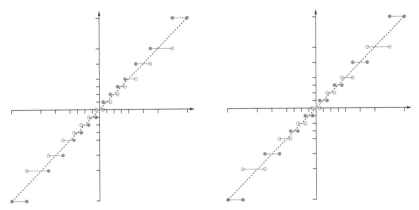

Figure 1.5 Round to nearest: (a) biased $\Box_n$; (b) unbiased $\Box$.

respectively. Note that, by Lemma 1.10, if $x \notin \mathbb{F}^*$ then *exactly* one of the integers a_{p-1} and b_{p-1} is even. We define the *round to nearest even* scheme by

$$x > 0 \Rightarrow \Box(x) = \begin{cases} \nabla(x), & \text{if } x \in [\nabla(x), \mu), \text{ or if } x = \mu \text{ and } a_{p-1} \text{ is even,} \\ \triangle(x), & \text{if } x \in (\mu, \triangle(x)], \text{ or if } x = \mu \text{ and } b_{p-1} \text{ is even,} \end{cases}$$

(1.7)

$$x < 0 \Rightarrow \Box(x) = -\Box(-x).$$

Note that this definition evens out the probability of rounding upward or downward; the rounding is *unbiased* (see Figure 1.5b).

When rounding a real number x of very large magnitude ($|x| > N_{max}^n$) it should be pointed out that despite the name *round to nearest*, x is actually rounded to $\text{sign}(x)\infty$, while $\text{sign}(x)N_{max}^n$ actually is the nearest element in $\mathbb{F}^*$.

As a final remark, it is clear that round to nearest *odd* can be defined in a completely analogous manner. One may then wonder whether the choice between rounding to nearest even or to nearest odd is relevant. The answer is, somewhat surprisingly, yes. This is demonstrated in the following example.

Example 1.3.1 *Consider the following scenario: let $\beta = 10$, $x = 0.45$, and suppose that we want to round x to two digits, after which we continue to round the result to one digit. Using round to nearest even produces $0.45 \rightarrow 0.4 \rightarrow 0 = \tilde{x}_e$, whereas round to nearest odd produces $0.45 \rightarrow 0.5 \rightarrow 1 = \tilde{x}_o$, which is not the nearest single-digit number seeing that $|x - \tilde{x}_e| = 0.45 < 0.55 = |x - \tilde{x}_o|$.*

Now, suppose that $\beta = 4$, $x = (0.22)_4$, and suppose once again that we want to round x to two digits, after which we continue to round the result to one digit. Using round to nearest even produces $(0.22)_4 \rightarrow (0.2)_4 \rightarrow (0)_4 = \tilde{x}_e$, whereas round to nearest odd produces $(0.22)_4 \rightarrow (0.3)_4 \rightarrow (1)_4 = \tilde{x}_o$, which now is the nearest single-digit number seeing that $|x - \tilde{x}_e| = (0.22)_4 = 5/8 > 3/8 = (0.21)_4 = |x - \tilde{x}_o|$.

This example illustrates the fact that when using an even base β, with $\beta/2$ odd, the best choice is round to nearest *even*. For an even base β, with $\beta/2$ even, however, the best choice is round to nearest *odd*. If the base itself β is odd, we can never have a tie (at least not in finite precision), so the choice of rounding never arises.

In particular, this means that for the popular choices $\beta = 10$ or 2, we should use round to nearest *even*, whereas for $\beta = 16$, round to nearest *odd* is the superior choice.

1.3.4 Rounding Errors

In the normalized range, the error produced when rounding a real number to a floating point system can be bounded in terms of the base β and precision p. We have the following bounds on the relative and absolute rounding errors.

THEOREM 1.13. *If x is a real number in the normalized range of $\mathbb{F} = \mathbb{F}_{\beta,p}$, then the relative error caused by rounding is bounded by $\varepsilon_M = \beta^{-(p-1)}$:*

$$\left| \frac{x - \bigcirc(x)}{x} \right| < \varepsilon_M.$$

Equivalently, the corresponding absolute error is bounded by $|x|\varepsilon_M$:

$$|x - \bigcirc(x)| < |x|\varepsilon_M.$$

The number $\varepsilon_M = \beta^{-(p-1)}$ is called the *machine epsilon* and is a very useful quantity in numerical error analysis. It is the distance between 1 and the next larger floating point number. We will encounter ε_M on several occasions throughout this text.

Proof. Without loss of generality, we may assume that x is positive. If x happens to be a member of $\mathbb{F}$, then there is no rounding error, and the claim follows trivially. Thus we only need to consider the case $x \notin \mathbb{F}$. Since x is in the normalized range, it has a representation $x = (b_0.b_1b_2 \ldots b_{p-1}b_p \ldots)_\beta \times \beta^e$, where $b_0 \neq 0$. Using the fact that $x \notin \mathbb{F}$, it follows that its nearest neighbors in $\mathbb{F}$, $\triangledown(x) = (b_0.b_1b_2 \ldots b_{p-1})_\beta \times \beta^e$ and $\triangle(x) = [(b_0.b_1b_2 \ldots b_{p-1})_\beta + \beta^{-(p-1)}] \times \beta^e$, are separated by the distance given by $\triangle(x) - \triangledown(x) = \beta^{-(p-1)} \times \beta^e = \varepsilon_M\beta^e$ and so $|x - \bigcirc(x)| < \varepsilon_M\beta^e$. Since x is normalized, we have that $x \geq 1 \times \beta^e$, so $|x - \bigcirc(x)| < x\varepsilon_M$. This gives the absolute error bound from which the relative bound follows immediately. Negative numbers are treated completely analogously. □

Note that in the case of rounding to nearest, the bounds of Theorem 1.13 can be decreased by a factor 0.5.

Exercise 1.14. *Show that the spacing between two adjacent floating point numbers x and y in the normalized range is bounded between $|x|\varepsilon_M/\beta$ and $|x|\varepsilon_M$.*

Example 1.3.2 *With base $\beta = 2$ and precision $p = 14$, the correctly rounded fraction $1/10$ is represented as:*

$$\triangledown(1/10) = (1.1001100110011)_2 \times 2^{-4},$$
$$\triangle(1/10) = (1.1001100110100)_2 \times 2^{-4}.$$

Thus we have

$$\left|\frac{1/10 - \triangledown(1/10)}{1/10}\right|$$

$$= \left|\frac{(1.10011001100110011\ldots)_2 \times 2^{-4} - (1.1001100110011)_2 \times 2^{-4}}{(1.10011001100110011\ldots)_2 \times 2^{-4}}\right|$$

$$= \left|\frac{(1.10011001100110011\ldots)_2 \times 2^{-20}}{(1.10011001100110011\ldots)_2 \times 2^{-4}}\right| = 2^{-16}.$$

$$\left|\frac{1/10 - \triangle(1/10)}{1/10}\right|$$

$$= \left|\frac{(1.10011001100110011\ldots)_2 \times 2^{-4} - (1.1001100110100)_2 \times 2^{-4}}{(1.10011001100110011\ldots)_2 \times 2^{-4}}\right|$$

$$= \left|\frac{(0.11001100110011001\ldots)_2 \times 2^{-15} - (1.0000000000000)_2 \times 2^{-15}}{(1.10011001100110011\ldots)_2 \times 2^{-4}}\right|$$

$$< \left|\frac{(011)_2 \times 2^{-17} - (100)_2 \times 2^{-17}}{(1.100)_2 \times 2^{-4}}\right| = \left|\frac{3 \times 2^{-17} - 4 \times 2^{-17}}{(1.100)_2 \times 2^{-4}}\right|$$

$$= \left|\frac{-1 \times 2^{-17}}{(1.100)_2 \times 2^{-4}}\right| = \frac{2}{3} \times 2^{-13}.$$

Both relative errors are clearly bounded by $\varepsilon_M = 2^{-13}$.

It is important to keep in mind that Theorem 1.13 does *not* hold for subnormal numbers. In this situation, we can no longer use the fact that the leading digit b_0 in the floating point representation of x is non-zero. This prevents us from obtaining the desired bounds. Nevertheless, a simple modification of the proof of Theorem 1.13 gives the following error bounds.

COROLLARY 1.15. *If x is a real number such that* $|x| = (b_0.b_1b_2 \ldots b_{p-1}b_p \ldots)_\beta \times \beta^{\check{e}}$ *with $b_i = 0$ for all $0 \le i < k \le p-1$ and $b_k \neq 0$, then the relative error caused by rounding to* $\mathbb{F}_{\beta,p}^{\check{e},\hat{e}}$ *is bounded by*

$$\left|\frac{x - \bigcirc(x)}{x}\right| < \beta^{-(p-1-k)}.$$

Equivalently, the corresponding absolute error is bounded by

$$|x - \bigcirc(x)| < |x|\beta^{-(p-1-k)}.$$

Thus, rounding in the subnormal range ($N_{min}^s \le |x| \le N_{max}^s$) leads to larger relative errors. The alternative, that is, having no subnormal numbers at all, would result in flushing any number with $|x| < N_{min}^n$ to zero, which is of course even less desirable. Note that the requirement $k \le p - 1$ ensures that $|x|$ is not smaller than the smallest positive subnormal number. Were this the case, the real number $|x|$ could be rounded *down* to zero, yielding a relative error of 1. It could also be rounded *up* to the smallest subnormal number, in which case the relative error could be arbitrarily large.

Example 1.3.3 *In the floating point system* $\mathbb{F}_{2,10}^{-5,5}$, *the correctly rounded real number* 10^{-100} *is represented as:*

$$\nabla(10^{-100}) = (0.000000000)_2 \times 2^{-5},$$
$$\triangle(10^{-100}) = (0.000000001)_2 \times 2^{-5}.$$

Thus we have the relative error bounds:

$$\left| \frac{10^{-100} - \nabla(10^{-100})}{10^{-100}} \right| = \left| \frac{10^{-100} - (0.000000000)_2 \times 2^{-5}}{10^{-100}} \right| = 1.$$

$$\left| \frac{10^{-100} - \triangle(10^{-100})}{10^{-100}} \right| = \left| \frac{10^{-100} - (0.000000001)_2 \times 2^{-5}}{10^{-100}} \right|$$

$$= \left| \frac{10^{-100} - 2^{-14}}{10^{-100}} \right| > \frac{10^{-5}}{10^{-100}} = 10^{95}.$$

1.4 FLOATING POINT ARITHMETIC

The main mathematical concern about computing over a set of floating point numbers (i.e., a set $\mathbb{F}$ of type $\mathbb{F}_{\beta,p}^{\check{e},\hat{e}}$ or $\mathbb{F}_{\beta,p}$) is that it is not arithmetically closed. This means that if we take two floating point numbers $x, y \in \mathbb{F}$ and choose an arithmetic operator $\star \in \{+, -, \times, \div\}$ then, in general, the result will not be exactly representable in the floating point system: $x \star y \notin \mathbb{F}$. This is a property that is *not* shared by the real numbers $\mathbb{R}$, nor even the rationals $\mathbb{Q}$. No matter how high precision we use, this problem remains.

The only way we can define arithmetic on a set of floating point numbers $\mathbb{F}$ then is to associate the exact (in $\mathbb{R}$) outcome of a floating point operation with a representable floating point number. Naturally, this can be achieved by rounding the exact result from $\mathbb{R}$ to $\mathbb{F}$. Given any one of the arithmetic operations $\star \in \{+, -, \times, \div\}$, let $\circledast \in \{\oplus, \ominus, \otimes, \oslash\}$ denote the corresponding operation carried out in $\mathbb{F}$. We say that the floating point arithmetic is of *maximum quality* if

(R5) $x, y \in \mathbb{F}$ and $\star \in \{+, -, \times, \div\} \Rightarrow x \circledast y = \bigcirc(x \star y).$

Property (R5) states that floating point arithmetic should yield the same result as though the computation were carried out with infinite precision, after which the exact result is rounded to the appropriate neighboring floating point. Thus the only error is incurred by the final rounding to $\mathbb{F}$, and consequently, the result of any arithmetic operation has the same quality as the rounding itself. It is true, but not obvious, that this can be practically implemented.[6]

---

[6]In fact, it suffices to use registers having $p + 2$ digits of precision, combined with a so-called *sticky bit* to obtain maximum quality in the arithmetic operations. See, for example, [Go91], [Kn98], or [Ko02].

THEOREM 1.16. *Let* $\star$ *denote one of the arithmetic operators* $+, -, \times, \div$. *Then, if* x *and* y *are normal floating point numbers with* $x \star y \neq 0$, *and* $x \circledast y$ *neither under- or overflowing, the relative error of the floating point operation is bounded by*

$$\left| \frac{x \star y - x \circledast y}{x \star y} \right| < \varepsilon_M.$$

Equivalently, the corresponding absolute error is bounded by

$$|x \star y - x \circledast y| < |x \star y| \varepsilon_M.$$

It is important to realize that Theorem 1.16 is only valid for *one single* floating point operation. All bets are off when several operations are involved. An expression like

$$f(x) = 1 \ominus ((1 \oplus x) \ominus 1)$$

may return a grossly incorrect value when $|x| < \varepsilon_M$, depending on the rounding mode. The following example serves as an illustration as to how these types of inaccuracies may seriously affect the outcome of a simple numerical experiment.

Example 1.4.1 *Consider the ternary shift map* $f : [0, 1] \rightarrow [0, 1]$ *defined by* $f(x) = 3x \bmod 1$. *This map has a period-4 cycle* $\frac{1}{10} \rightarrow \frac{3}{10} \rightarrow \frac{9}{10} \rightarrow \frac{7}{10} \rightarrow \frac{1}{10}$. *Starting a numerical iteration (over* $\mathbb{F}_{2,53}$) *at* $x_0 = \frac{1}{10}$, *however, produces the following orbit:*

$$x(0) = 0.10000000000000001 \quad x(1) = 0.30000000000000004$$
$$x(2) = 0.90000000000000013 \quad x(3) = 0.70000000000000018$$
$$x(4) = 0.10000000000000053 \quad x(5) = 0.3000000000000016$$

$$\vdots$$

$$x(47) = 0.15273362128584456 \quad x(48) = 0.45820086385753367$$
$$x(49) = 0.37460259157260101 \quad x(50) = 0.12380777471780302$$
$$x(51) = 0.37142332415340906 \quad x(52) = 0.11426997246022719$$

After less than 50 iterates, there is no sign of the periodic orbit. The reason is that the function f *is expanding, that is, its derivate (when defined) is greater than 1 everywhere. As a consequence, even very small initial errors will eventually be grossly magnified and completely saturate the computed orbit. This intrinsic property of expanding maps makes the numerical study of chaotic dynamical systems a very challenging topic.*

Another consequence of computing with finite precision is that many basic mathematical laws no longer hold. For example, addition and multiplication are no longer associative.

Example 1.4.2 *Consider the numbers* $x = 1.234 \times 10^4$, $y = -1.235 \times 10^4$, *and* $z = 1.002 \times 10^1$, *all belonging to* $\mathbb{F}_{10,4}^{-9,9}$. *Rounding to nearest even, we have*

$$(x \oplus y) \oplus z = -1.000 \times 10^1 \oplus 1.002 \times 10^1 = 2.000 \times 10^{-2},$$

whereas

$$x \oplus (y \oplus z) = 1.234 \times 10^4 \ominus 1.234 \times 10^4 = 0.000 \times 10^{-9}.$$

The first result is exact, while the second suffers from inaccuracies caused by a lack of precision.

Exercise 1.17. *How many pairs of floating point numbers can be exactly added in the set $\mathbb{F}_{2,3}^{-1,2}$? How many pairs cannot? What about the general case $\mathbb{F}_{\beta,p}^{\check{e},\hat{e}}$?*

Exercise 1.18. *Assuming the property (R5), show that the following statements always hold true:*

(1) $x \in \mathbb{F} \Rightarrow 1 \otimes x = x$;

(2) $x \in \mathbb{F} \setminus \{-\infty, 0, +\infty\} \Rightarrow x \oslash x = 1$;

(3) $x \in \mathbb{F}$ (with $\beta = 2$) $\Rightarrow 0.5 \otimes x = x \oslash 2$.

(4) $x, y \in \mathbb{F} \Rightarrow (x \ominus y = 0) \Rightarrow (x = y)$.

Also show that statement (4) is false if $\mathbb{F}$ has no subnormal numbers.

In view of (R5), we can give a computational definition of the machine epsilon. Even if the base and precision of the underlying floating point system are unknown, this definition allows for a direct computation of ε_M.

DEFINITION 1.19. *In a floating point system with base β and precision p, we call the number $\varepsilon_M = \beta^{-(p-1)}$ the machine epsilon. It is the smallest positive floating point number x that satisfies $1 < 1 \oplus x$ when rounding down, that is, $\varepsilon_M = \min\{x \in \mathbb{F} : 1 < \bigtriangledown(1 \oplus x)\}$.*

Note that the condition $1 < \bigtriangledown(1 \oplus x)$ is very different from the *seemingly* equivalent condition $0 < \bigtriangledown(x)$. Some aggressively optimizing compilers do not realize this distinction and thus produce grossly incorrect code. The smallest floating point number x satisfying $0 < \bigtriangledown(x)$ is called the *machine eta* and is denoted by η_M. This is the smallest positive subnormal number and is thus equal to $\beta^{\check{e}-p+1} = \beta^{\check{e}} \varepsilon_M$. In Problem 2 of Computer Lab I the reader is asked to write a small program that computes both ε_M and η_M. These computations can safely be carried out in the default rounding mode, round to nearest even, since the least significant bit of both 1.0 and 0.0 is zero, which is an even number.

1.5 THE IEEE STANDARD

In the early days of computing, each brand of computer had its own implementation for floating point operations. This had the unfortunate effect that the outcome of a computation heavily depended on the precise type of machine used to perform the calculations. Naturally, this also severely limited the portability of programs, as code that worked perfectly well on one machine could crash on another. Even worse, most computer manufacturers had their own internal *representation* of floating point numbers. Of course, this meant that data transfer between different machines became a highly complex task.

In the second half of the 1980s, an international standard for the use and representation of floating point numbers was agreed upon. The standard is actually several standards: the IEEE p754, released in 1985, which deals exclusively with binary representations, and the IEEE p854, released in 1987, which in a common framework considers both base 2 and 10 (see [IE85] and [IE87], respectively). Recently, both of these documents have been superseded by the IEEE-754-2008 (see [IE08]). As this standard has yet to be adhered to by hardware manufacturers, we will focus on the two older standards when referring to the IEEE standard.

Besides demanding maximal quality[7] of the arithmetic of floating point numbers, the standard also requires a consistent treatment of exceptions (e.g., division by zero), which greatly facilitates the construction of robust code. The IEEE standard also requires the presence of the four basic rounding modes: round up, round down, round to zero, and round to nearest (even).

For two very nice expositions of the IEEE floating point standard, see [Go91] and [Ov01].

1.5.1 The IEEE Formats

The IEEE standard specifies two basic types of floating point numbers: the single and double formats. Extended versions of these formats are also mentioned, although only minimal requirements (as opposed to exact descriptions) for these are specified. Whereas the extended double format is provided on most architectures, the extended single format has found little support. We will therefore only cover the extended double format, which henceforth is denoted extended.

The standard also introduces three special symbols: -Inf, +Inf, and NaN. The first two are direct analogues to the mathematical notion of $\pm\infty$. Any finite floating point number x satisfies $-\text{Inf} < x < +\text{Inf}$. A real number greater than the largest finite floating point number is represented as either N^n_{max} or +Inf, depending on the rounding mode. Similarly, a real number smaller than the smallest finite floating point number is represented as either $-N^n_{max}$ or -Inf. The symbol NaN, which is short for *not a number*, is returned whenever the outcome of a floating point operation is undefined, for example, $0/0$ or Inf-Inf. The IEEE standard will in some cases distinguish between -0 and $+0$, which can lead to some unexpected behavior. We will return to this matter in Section 2.3.3.

To simplify the exposition, in all that follows we will consider only floating point representations in binary base. Although this base has some special advantages, the theory presented can easily be generalized to a setting with an arbitrary base.

On almost all commercial computers, the two basic formats single and double are implemented using exactly the same mold, varying only two parameters. Each format is made up of a fixed number $F_\star$ of bits (binary integers),

---

[7]The IEEE standard requires that the basic operations $\{+, -, \times, \div\}$ and $\sqrt{}$ return the floating point nearest to the exact result with regard to the rounding mode. Rather surprisingly, no such demands are imposed on the trigonometric and exponential functions. On a HP 9000/700, the argument $x = 2.50 \times 10^{17}$ produces the grossly incorrect $\sin x = 4.14415 \times 10^7$.

where the subscript $\star \in \{s, d\}$ indicates the specific format in question. The first bit encodes the sign of the floating point number, the following $E_\star$ bits correspond to the exponent, and the remaining $M_\star$ bits represent the mantissa. Thus, any two elements of $\{F_\star, E_\star, M_\star\}$ completely define the format in question. Actually, each format has precision $M_\star + 1$. This is achieved by a *hidden bit*: since a normal floating point number represented in base 2 must start with a one, there is no need to explicitly store this leading bit. Note that this trick only works for binary representations. The exponent also has a little twist to it: we are not storing the actual exponent but rather a *biased* version of it. Using $E_\star$ bits, we can represent any integer between 0 and $2^{E_\star} - 1$. We form the actual exponent by subtracting the bias $B_\star = 2^{E_\star - 1} - 1$ from the stored number. This gives a exponent range of $[-2^{E_\star - 1} + 1, 2^{E_\star - 1}]$. The two boundary points, however, are reserved for non-normal numbers.

Consider the $F_\star$-bit string $[\sigma; e_1 e_2 \ldots e_{E_\star}; m_1 m_2 \ldots m_{M_\star}]$, and let $E = (e_1 e_2 \ldots e_{E_\star})_2$ and $M = (0.m_1 m_2 \ldots m_{M_\star})_2$. Then the floating point number x represented by the string is decoded as follows:

(a) if $E = 2^{E_\star} - 1$ and $M \neq 0$, then $x = \text{NaN}$;
(b) if $E = 2^{E_\star} - 1$ and $M = 0$, then $x = (-1)^\sigma \text{Inf}$;
(c) if $0 < E < 2^{E_\star} - 1$, then $x = (-1)^\sigma 1.M \times 2^{E - B_\star}$;
(d) if $E = 0$ and $M \neq 0$, then $x = (-1)^\sigma 0.M \times 2^{1 - B_\star}$;
(e) if $E = 0$ and $M = 0$, then $x = (-1)^\sigma \times 0$.

We see that case (c) corresponds to the set of normal numbers, whereas case (d) deals with the subnormal numbers. Note that cases (d) and (e) could be merged, although we chose not to do so seeing that zero is not a subnormal number.

1	$E_\star$	$M_\star$
σ	e	m

Figure 1.6 The basic IEEE formats.

The `single` format consists of 32 bits, of which 8 bits correspond to the exponent and the remaining 23 bits represent the mantissa. In other words, we have $\{F_s, E_s, M_s\} = \{32, 8, 23\}$. The smallest and largest positive normal numbers in `single` format are seen to be $N^n_{min} = 2^{-126} \approx 1.2 \times 10^{-38}$ and $N^n_{max} = (2 - 2^{-23}) \times 2^{127} \approx 2^{128} \approx 3.4 \times 10^{38}$, respectively. The machine epsilon is $\varepsilon_M = 2^{-23} \approx 1.2 \times 10^{-7}$. This means that we should expect about seven significant decimal digits.

1	8	23
σ	e	m

Figure 1.7 The IEEE `single` format.

The `double` format consists of 64 bits, of which 11 bits correspond to the exponent and the remaining 52 bits represent the mantissa, that is, we have $\{F_d, E_d, M_d\} = \{64, 11, 52\}$. The smallest and largest positive normal numbers in `double` format are seen to be $N^n_{min} = 2^{-1022} \approx 2.2 \times 10^{-308}$ and $N^n_{max} = (2 - 2^{-52}) \times 2^{1023} \approx 2^{1024} \approx 1.8 \times 10^{308}$, respectively. The machine epsilon is

Table 1.1 The most common IEEE formats

	single	double	extended
Format width in bits	32	64	≥ 79
Exponent width in bits	8	11	≥ 15
Precision p	24	53	≥ 64
Exponent bias	$+127$	$+1023$	unspecified
Maximal exponent	$+127$	$+1023$	$\geq +16383$
Minimal exponent	-126	-1022	≤ -16382

$\varepsilon_M = 2^{-52} \approx 2.2 \times 10^{-16}$, so we can expect about 16 significant decimal digits from the double format.

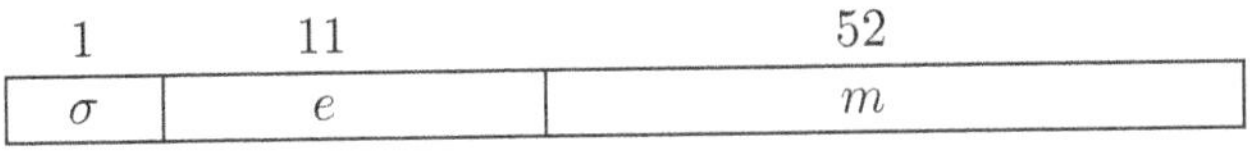

1	11	52
σ	e	m

Figure 1.8 The IEEE double format.

1.5.2 The *extended* Format

In contrast to the single and double formats, the IEEE standard does not specify absolute parameters for the extended format. Instead, minimal requirements are given, and it is up to each individual manufacturer to decide the precise parameters to be used. The requirements for the three formats are illustrated below.

Considering that the extended format is not precisely specified, it is rather unfortunate that it has become the most commonly used format. To make matters worse, the non-expert user is often unaware of this fact. The reason for this state of affairs is that almost all computers perform intermediate computations in the widest registers available to them. Even the simplest computation, involving only two double type variables, will be converted to and performed in extended format, after which the result is rounded back to the double format. This can (and often does) lead to quite unexpected results, which are very hard to "debug" (see Example 1.6.1). In light of this, we will spend quite some time describing the various "flavors" of the extended format.

Let us begin with the 128-bit extended format, which is provided on the SPARC architecture. This format follows the generic mold described in the previous section, with parameters $\{F_e, E_e, M_e\} = \{128, 15, 112\}$. Thus, the smallest and largest positive normal numbers in the 128-bit extended format are seen to be $N^n_{min} = 2^{-16382} \approx 3.4 \times 10^{-4932}$ and $N^n_{max} = (2 - 2^{-112}) \times 2^{16383} \approx 2^{16384} \approx 1.2 \times 10^{4932}$, respectively. The machine epsilon is $\varepsilon_M = 2^{-112} \approx 1.9 \times 10^{-34}$. This means that we should expect about 34 significant decimal digits from the 128-bit extended format.

1	15	112
σ	e	m

Figure 1.9 The SPARC 128-bit extended format.

In contrast to SPARC, the Intel x86 and Pentium architectures provide an 80-bit extended format, made up by ten 8-bit bytes.[8] This format consists of a 1-bit sign, a 15-bit (biased) exponent, and a 64-bit mantissa. In the 64-bit mantissa no hidden bit is employed, which leads to some minor peculiarities.

Consider the 80-bit string $[\sigma; e_1 e_2 \ldots e_{15}; m_0 m_1 m_2 \ldots m_{63}]$. Let $E = (e_1 e_2 \ldots e_{15})_2$ and $M = (0.m_1 m_2 \ldots m_{63})_2$. Then the floating point number x represented by the string is decoded as follows:

(a) if $m_0 = 1$, $E = 32767$, and $M \neq 0$, then $x = \text{NaN}$;
(b) if $m_0 = 1$, $E = 32767$, and $M = 0$, then $x = (-1)^\sigma \text{Inf}$;
(c) if $m_0 = 1$ and $0 < E < 32767$, then $x = (-1)^\sigma 1.M \times 2^{E-16383}$;
(d) if $m_0 = 0$, $E = 0$, and $M \neq 0$, then $x = (-1)^\sigma 0.M \times 2^{-16382}$;
(e) if $m_0 = 0$, $E = 0$, and $M = 0$, then $x = (-1)^\sigma \times 0$;
(f) if $m_0 = 0$ and $0 < E < 32767$, then x is not defined;
(g) if $m_0 = 1$ and $E = 0$, then $x = (-1)^\sigma 1.M \times 2^{-16382}$.

The numbers corresponding to case (g) are called *pseudo-subnormal numbers*. These are never generated as results but may appear as operands, in which case they are implicitly converted to the corresponding normal numbers as in (c).

The smallest and largest positive normal numbers in the 80-bit extended format are seen to be $N^n_{min} = 2^{-16382} \approx 3.4 \times 10^{-4932}$ and $N^n_{max} = (2 - 2^{-63}) \times 2^{16383} \approx 2^{16384} \approx 1.2 \times 10^{4932}$, respectively. The machine epsilon is $\varepsilon_M = 2^{-63} \approx 1.1 \times 10^{-19}$. This means that we should expect about 19 significant decimal digits from the 80-bit extended format.

1	15	64
σ	e	m

Figure 1.10 The Intel 80-bit extended format.

The IBM S/390 G5 series, which uses a hexadecimal base, has hardware support for the IEEE formats with parameters matching those of the SPARC (see [SK99]). The CRAY T90 series has also moved toward the IEEE formats but with some important exceptions. First, there is a slight linguistic discrepancy: the CRAY single format is 64 bits wide and thus corresponds to the IEEE double. Similarly, the CRAY double format is 128 bits wide and therefore corresponds to the IEEE extended. What is more serious is that the CRAY architecture does not treat subnormal numbers according to the IEEE standard. In fact, all subnormal numbers are forced to zero (see [Ga96]). As we have seen, this leads to the violation of several important mathematical laws. To further confuse matters, the Macintosh PowerPC Numerics Environment provides a *double-double* format that is 128 bits wide. The exponent field, however, is only 11 bits wide, which is lower than the requirement for an IEEE extended format. The *double-double* is implemented in software combining two double formats in a quite complicated manner.

In Table 1.2, we list some important facts about the various formats. Here, m-bits denotes the number of bits reserved for the mantissa, and e-bits corresponds to the exponent field.

---

[8]Under the UNIX System V operating system, however, the format is made up by three 32-bit words, leaving the 16 highest addressed bits unused.

Table 1.2 Summary of the most common IEEE formats

format	bits	m-bits	e-bits	N_{min}^s	N_{max}^n
single	32	23 + 1	8	1.4×10^{-45}	3.4×10^{38}
double	64	52 + 1	11	4.9×10^{-324}	1.8×10^{308}
INTEL extended	80	64	15	3.6×10^{-4951}	1.2×10^{4932}
SPARC extended	128	112 + 1	15	6.5×10^{-4966}	1.2×10^{4932}

1.6 EXAMPLES OF FLOATING POINT COMPUTATIONS

A common way to determine the accuracy of a specific computation is to gradually increase the precision until the result stabilizes. If adding more bits to the floating point representation does not alter the result of the computation, one usually accepts the result as being correct. The following example illustrates the false sense of security given by this approach.

Example 1.6.1 *Consider the function*

$$f(x, y) = 333.75y^6 + x^2(11x^2y^2 - y^6 - 121y^4 - 2) + 5.5y^8 + x/(2y).$$

As observed in [Ru88], using FORTRAN on an IBM S/370 ($\beta = 16$), the function evaluated at the point $(\tilde{x}, \tilde{y}) = (77617, 33096)$ produces the following output:

type	p	$f(\tilde{x}, \tilde{y})$
REAL∗4	24	1.172603...
REAL∗8	53	1.1726039400531...
REAL∗10	64	1.172603940053178...

Using C or C++ (with gcc/g++) on an Intel Pentium III chip ($\beta = 2$), we get

type	p	$f(\tilde{x}, \tilde{y})$
float	24	178702833214061281280
double	53	178702833214061281280
long double	64	178702833214061281280

Using C or C++ (with gcc/g++) on a Sun UltraSPARC ($\beta = 2$), we get

type	p	$f(\tilde{x}, \tilde{y})$
float	24	257178416384078908222768939008
double	53	1.17260394005317869492444406059...
long double	113	1.17260394005317869492444406059...

Although all coefficients are exactly representable in base 2 (and thus in base 16), the rounding errors render the result useless. The correct answer is actually $-0.8273960599...$, which means that we did not even get the sign right.

These discrepancies are due to the fact that the two terms $T_1 = 5.5\tilde{y}^8$ and $T_2 = 333.75\tilde{y}^6 + \tilde{x}^2(11\tilde{x}^2\tilde{y}^2 - \tilde{y}^6 - 121\tilde{y}^4 - 2)$ are very large in modulus and almost cancel:

$$T_1 = +7917111340668961361101134701524942848$$

$$T_2 = -7917111340668961361101134701524942850.$$

Since the sum of these terms is $T_1 + T_2 = -2$, we are left with just

$$f(\tilde{x}, \tilde{y}) = T_1 + T_2 + \tilde{x}/(2\tilde{y}) = -2 + \tilde{x}/(2\tilde{y}),$$

which gives

$$f(\tilde{x}, \tilde{y}) = -2 + \frac{77617}{2 \times 33096} \approx -0.8273960599.$$

Of course, when computing with any of the above-mentioned floating points formats, we have cancellation, and the sum of the two huge terms (both of magnitude $\approx 7.9 \times 10^{36}$) is evaluated as zero. This results in the approximate function value

$$f(\tilde{x}, \tilde{y}) = 0 + \frac{77617}{2 \times 33096} \approx 1.1726039400531.$$

This, however, does not explain the results from the Intel Pentium III chip. In this case, with no explicit instructions passed on to the compiler, all intermediate results are converted and worked upon in the long double *format by default.*

For a more thorough analysis of this example, see [EW02] and [CV01].

The next example deals with the (seemingly) simple task of generating the graph of a polynomial.

Example 1.6.2 *Plot the graph, and search for roots of the polynomial*

$$p(t) = t^6 - 6t^5 + 15t^4 - 20t^3 + 15t^2 - 6t + 1.$$

MATLAB *produces the non-smooth graph illustrated in Figure 1.11. This picture is clearly wrong: a polynomial of degree n can have at most $n - 1$ local extrema. Even if we minimize the number of floating point operations by evaluating the polynomial via Horner's method*

$$p(t) = ((((((t - 6)t + 15)t - 20)t + 15)t - 6)t + 1),$$

the resulting graph is clearly not correct. Note, however, that we can factor the polynomial as $p(t) = (t - 1)^6$, which is (correctly) plotted as the smooth graph in Figure 1.11. It follows that the graph should lie above the t-axis, except at the multiple root $t^* = 1$.

The reason why the computed values approximate the graph so poorly is that the condition number[9] of $p(t)$ near $t^* = 1$ is very large. Without going into explicit

---

[9]For functions, the condition number is the maximal value of the ratio between the relative errors in the function and the relative errors in the argument:

$$\kappa(t, t^*) = \left| \frac{f(t) - f(t^*)}{f(t)} \right| \div \left| \frac{t - t^*}{t} \right|.$$

In the case of a differentiable function, we can let t tend to t^* in the first factor, producing

$$\kappa(t, t^*) \approx |f'(t^*)| \div \left| \frac{t - t^*}{t} \right|.$$

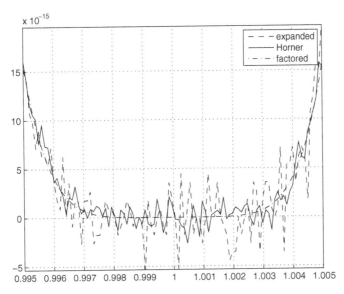

Figure 1.11 A smooth graph of a polynomial?

calculations, the condition number of p near t^* is given by

$$\kappa(t, t^\star) \approx 6 \left| \frac{t}{t - t^\star} \right|.$$

As t approaches the multiple root at t^*, we see that κ tends to $+\infty$. A large condition number translates to poor accuracy, as is illustrated in Figure 1.11. Note that this situation would not occur if the multiple root was positioned at 0 rather than 1.

The conclusion we draw is that function evaluations depend on the function *representation* when computed over $\mathbb{F}$. This is a difficult fact to accept for most mathematicians, who are used to computing over $\mathbb{R}$.

After these unnerving examples, let us show how one can obtain rigorous mathematical statements using the computer. By utilizing the directed rounding modes, we can enclose the exact result of certain computations. These techniques will be generalized and studied in detail in Chapter 2.

Example 1.6.3 *It is well-known that the infinite series $S = \sum_{k=1}^{\infty} k^{-2}$ has the exact value $\pi^2/6$. Assume for the moment that we are unaware of this, and suppose that we need to find an approximation to S, say to 12 decimal places. Clearly, we cannot sum an infinite number of terms on the computer, so let us split the series into two pieces:*

$$S = \sum_{k=1}^{\infty} k^{-2} = \sum_{k=1}^{N} k^{-2} + \sum_{k=N+1}^{\infty} k^{-2} = S_N + S_N^\star.$$

Our strategy is now to bound the infinite part $S_N^\star$ by mathematical means, whereas we calculate the finite part S_N using the computer.

In order to achieve 12 correct decimal places, we must ensure that the upper and lower bounds for $S_N^\star$ differ by at most 5×10^{-13}. By a simple geometric argument (draw the picture), we have

$$\int_{N+1}^{\infty} \frac{dx}{x^2} < S_N^\star < \int_{N+1}^{\infty} \frac{dx}{(x-1)^2},$$

which produces the bounds $\frac{1}{N+1} < S_N^\star < \frac{1}{N}$ of width $\delta_N = \frac{1}{N(N+1)}$. Taking $N = 2 \times 10^6$ gives $\delta_N < 2.5 \times 10^{-13}$, which should do nicely, assuming that we can compute the finite part S_N accurately.

So let $N = 2 \times 10^6$, and compute the sum $S_N = \sum_{k=1}^{N} k^{-2}$ with double *precision, using the directed rounding modes. This produces the following output:*

```
Rounded down:  S_N = 1.644 933 566 626 364 25
Rounded up:    S_N = 1.644 933 567 070 448 80.
```

As is plain to see, the results differ already in the ninth decimal. Since all terms are positive, however, the IEEE standard guarantees that the exact result is bounded from below by the result obtained when always rounding down. Analogously, the exact result is bounded from above by the result obtained when always rounding up. Thus we can enclose the exact value of S_N in the interval

$$[1.64493356662636425, 1.64493356707044880] \overset{\text{def}}{=} 1.64493356^{707044880}_{662636425}.$$

As the number of terms to be summed is known in advance, we can get better accuracy by adding the terms in increasing order (can you explain why?). Doing so yields the results

```
Rounded down:  S_N = 1.644 933 566 848 350 46
Rounded up:    S_N = 1.644 933 566 848 352 46,
```

which now differ in the fifteenth decimal. Once again, we know that the exact value of S_N is contained in the interval

$$[1.64493356684835046, 1.64493356684835246] = 1.64493356684835^{246}_{046},$$

which allows us to refine our numerical enclosure of the partial sum:

$$S_N \in 1.64493356684835^{246}_{046} \subset 1.64493356^{707044880}_{662636425}.$$

Combining this information with the fact that $\frac{1}{N+1} < S_N^\star < \frac{1}{N}$ produces the final enclosure:

$$S \in 1.644934066848^{35253}_{10028},$$

which is correct to 12 decimal places. Compare this to the "exact" value $S = \pi^2/6 \approx 1.64493406684822630$, where we have approximated π by its 50 leading digits.

It is exactly this mixture of mathematics and properly rounded numerical computations that opens the door to *validated numerics*. As we have seen, these techniques allow us to construct computer-aided mathematical *proofs*, just like our recent proof that all digits of the approximation $S \approx 1.644934066848$ are correct.

1.7 COMPUTER LAB I

Problem 1. Write a program that computes the factorial $n! \overset{\text{def}}{=} 1 \cdot 2 \cdots n$ of a given integer n. Use your program to compute the 40 first factorials. Do you notice anything strange? If so, try to explain what is happening.

Problem 2. Write a program that computes the smallest positive machine representable number η_M, and the machine epsilon ε_M. What are the values you get? Try to print them in hexadecimal form, too (see the code segment in Listing 1.1).

Problem 3. (a) Find an IEEE double-precision floating point number $x \in (1, 2)$ such that $x \otimes \frac{1}{x} \neq 1$. (b) Find the smallest such number (possibly by a brute force search).

Problem 4. Define the function $f(x, y) = 9x^4 - y^4 + 2y^2$. Your objective is to compute $f(40545, 70226)$. Write a program that evaluates the function using each of the formats int, float, and double. What is the correct answer?

Problem 5. Write a program that switches the rounding mode on your computer. (Hint: for C/C++ programs, use the header file round.h listed in Listing 2.1. MATLAB programs can use the file setround.m from Listing 2.2.) Make your program compute $1/10$ in various rounding modes. Make sure you output enough decimals or, even better, print the results in hexadecimal.

Problem 6. Write a small interval arithmetic routine supporting the arithmetic operations with directed rounding. Use the routine to compute $F([1, 2])$ and $G([1, 2])$, where $f(x) = \frac{7x - (x+1)^2}{3x}$ and $g(x) = \frac{7x - (x+1)^2}{3x - 2}$, respectively.

Listing 1.1.   A simple Hex printer for C/C++ codes

```
1  /* A function for printing in Hex format. */
2  #include <iostream>
3  using namespace std;
4
5  #define HI_BITS 1 //Assumes "Little Endian" storage. Swap
6  #define LO_BITS 0 //these two on a "Big Endian" system.
7
8  void printHex(double x) {
9    cout << hex << ((int *) &x)[HI_BITS] << " "
10         << hex << ((int *) &x)[LO_BITS] << endl;
11 }
```

Chapter Two

Interval Arithmetic

IN THIS CHAPTER, we will briefly describe the fundamentals of interval arithmetic. We will also discuss how to implement the arithmetic in a programming environment.

Simply put, interval arithmetic is an arithmetic for *inequalities*. To illustrate this point, let us assume that we want to compute the area of a rectangle with side lengths ℓ_1 and ℓ_2. Given the measurements $\ell_1 = 10.3 \pm 0.1$ and $\ell_2 = 4.4 \pm 0.2$, what can we say about the area $A = \ell_1 \cdot \ell_2$? If we express our measurements in terms of the bounds $|\ell_1 - 10.3| \le 0.1$ and $|\ell_2 - 4.4| \le 0.2$, then (using the triangle inequality) all we can say is that $|\ell_1 \cdot \ell_2 - 10.3 \cdot 4.4| \le 0.2 \cdot 10.3 + 0.1 \cdot 4.4 + 0.1 \cdot 0.2$, that is, $|A - 45.32| \le 2.52$. If, on the other hand, we view the measurements as the inequalities $10.2 \le \ell_1 \le 10.4$ and $4.2 \le \ell_2 \le 4.6$, the optimal answer is obvious: the area must satisfy $42.84 = 10.2 \cdot 4.2 \le A \le 10.4 \cdot 4.6 = 47.84$, which translates into the slightly improved bound $|A - 45.34| \le 2.5$.

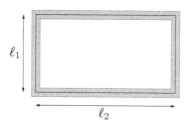

Figure 2.1 A rectangle with sides ℓ_1 and ℓ_2.

The calculations in the latter case can be summarized as a single multiplication of two intervals:

$$[10.2, 10.4] \times [4.2, 4.6] = [42.84, 47.84].$$

Interval arithmetic justifies this extension of the real arithmetic and provides an elegant means of computing with inequalities. For a concise reference on this topic, see [Mo66], [Mo79], [AH83], [Ne90], or [MK09]. Early references are [Yo31], [Dw51], [Wa56], [Su58], and [Mo59].

2.1 REAL INTERVALS

In what follows, our basic elements will be closed and bounded intervals of the real line. We will adopt the shorthand notation

$$\boldsymbol{a} = [\underline{a}, \overline{a}] = \{x \in \mathbb{R} : \underline{a} \le x \le \overline{a}\}$$

and consider the set of all such intervals of the real line:

$$\mathbb{IR} = \{[\underline{a}, \overline{a}] : \underline{a} \leq \overline{a}; \quad \underline{a}, \overline{a} \in \mathbb{R}\}.$$

Note that we allow for degenerate intervals $\boldsymbol{a}$ with $\underline{a} = \overline{a}$. We will refer to these intervals as being *thin*. A natural embedding of $\mathbb{IR}$ in $\mathbb{R}^2$ is given by the mapping $g \colon \mathbb{IR} \to \mathbb{R}^2$, defined by $[\underline{a}, \overline{a}] \mapsto (\underline{a}, \overline{a})$. Geometrically, this corresponds to viewing $\mathbb{IR}$ as the region in $\mathbb{R}^2$ above and on the diagonal $y = x$. Points in $\mathbb{R}^2$ lying on the diagonal correspond to thin intervals.

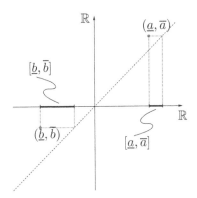

Figure 2.2 Identifying $\mathbb{IR}$ with $\{(x, y) \in \mathbb{R}^2 : y \geq x\}$.

Example 2.1.1 *The elements* $[-3, 4]$, $[1, 1]$, *and* $[\pi, 7]$ *all belong to* $\mathbb{IR}$, *whereas* $[2, -1]$ *and* $[-\infty, 0]$ *do not.*

Being sets, the elements of $\mathbb{IR}$ inherit the natural set relations, such as $=, \subseteq, \subset$, and $\overset{\circ}{\subset}$, defined by

$$\begin{aligned}
\boldsymbol{a} = \boldsymbol{b} \quad &\Leftrightarrow \quad \underline{a} = \underline{b} \text{ and } \overline{a} = \overline{b} \\
\boldsymbol{a} \subseteq \boldsymbol{b} \quad &\Leftrightarrow \quad \underline{b} \leq \underline{a} \text{ and } \overline{a} \leq \overline{b} \\
\boldsymbol{a} \subset \boldsymbol{b} \quad &\Leftrightarrow \quad \boldsymbol{a} \subseteq \boldsymbol{b} \text{ and } \boldsymbol{a} \neq \boldsymbol{b} \\
\boldsymbol{a} \overset{\circ}{\subset} \boldsymbol{b} \quad &\Leftrightarrow \quad \underline{b} < \underline{a} \text{ and } \overline{a} < \overline{b}
\end{aligned}$$

We can partially order[1] the set $\mathbb{IR}$ in several ways. Emphasizing the set-valued properties of $\mathbb{IR}$, we can use $\subseteq$ as a partial ordering. Preserving the natural ordering of the real numbers, we may extend the relation $\leq$ to mean

$$\boldsymbol{a} \leq \boldsymbol{b} \quad \Leftrightarrow \quad \underline{a} \leq \underline{b} \text{ and } \overline{a} \leq \overline{b}.$$

This also provides a partial ordering of $\mathbb{IR}$.

By somewhat abusing our interval notation, we often identify a real number a with the corresponding thin interval $[a, a]$. It then makes sense to define the relation

$$a \in \boldsymbol{b} \quad \Leftrightarrow \quad \underline{b} \leq a \text{ and } a \leq \overline{b},$$

---

[1] A relation $\sim$ is a *partial order* on a set S if, for all $a, b, c \in S$, it satisfies: (1) reflexivity: $a \sim a$; (2) antisymmetry: $a \sim b$ and $b \sim a$ implies $a = b$; and (3) transitivity: $a \sim b$ and $b \sim c$ implies $a \sim c$.

which is really a special case of $a \subseteq b$. All of these relations can be complemented by their logical opposites $\neq$, $\not\subseteq$, $\not\subset$, $\not\overset{\circ}{\subset}$, and $\not\ni$.

We can also equip $\mathbb{R}$ with analogues to the set operations $\cup$ and $\cap$. Both operations, however, require minor adjustments. First, taking the union of two intervals may not result in a new interval. To overcome this problem, we introduce the notion of forming the *hull* of two intervals:

$$a \sqcup b = [\min\{\underline{a}, \underline{b}\}, \max\{\overline{a}, \overline{b}\}].$$

It is clear that the resulting interval contains the union of a and b. Second, the intersection of two intervals a and b is empty if either $\overline{a} < \underline{b}$ or $\overline{b} < \underline{a}$. Because of this, we must add the empty set (denoted by $\emptyset$) to $\mathbb{R}$ for the intersection operator to be well-defined. When the intervals a and b have at least one point in common, the intersection is the standard one. Thus we have

$$a \cap b = \begin{cases} \emptyset & : \text{if } \overline{a} < \underline{b} \text{ or } \overline{b} < \underline{a}, \\ [\max\{\underline{a}, \underline{b}\}, \min\{\overline{a}, \overline{b}\}] & : \text{otherwise.} \end{cases}$$

Example 2.1.2 *Let* $a = [1, 3]$, $b = [1, \pi]$, $c = [-2.3, 4]$, *and* $d = [4, 5]$. *Then* $a \overset{\circ}{\subset} c$, $a \subset b$, $a \sqcup b = [1, \pi]$, $a \sqcup d = [1, 5]$, $a \cap d = \emptyset$, *and* $c \cap d = [4, 4]$.

Given an interval $a \in \mathbb{R}$, we define the following real-valued functions:

$$\text{rad}(a) = \tfrac{1}{2}(\overline{a} - \underline{a}) \qquad (\text{the } radius \text{ of } a),$$
$$\text{mid}(a) = \tfrac{1}{2}(\overline{a} + \underline{a}) \qquad (\text{the } midpoint \text{ of } a).$$

Thus we can write $a = [\text{mid}(a) - \text{rad}(a), \text{mid}(a) + \text{rad}(a)]$, and it follows that

$$\xi \in x \quad \Leftrightarrow \quad |\xi - \text{mid}(x)| \leq \text{rad}(x),$$

for any interval x. Two additional real-valued functions that often come in handy are

$$\text{mig}(a) = \min\{|a| : a \in a\} \qquad (\text{the } mignitude \text{ of } a),$$
$$\text{mag}(a) = \max\{|a| : a \in a\} \qquad (\text{the } magnitude \text{ of } a).$$

These functions provide us with the smallest and largest distance to the origin attained by elements of a. There are explicit, computable formulas for these functions:

$$\text{mig}(a) = \begin{cases} 0 & : \text{if } 0 \in a, \\ \min\{|\underline{a}|, |\overline{a}|\} & : \text{otherwise;} \end{cases} \qquad \text{mag}(a) = \max\{|\underline{a}|, |\overline{a}|\}.$$

Combining the two functions, we can form the *absolute value* of an interval:

$$\text{abs}(a) = \{|a| : a \in a\} = [\text{mig}(a), \text{mag}(a)].$$

In contrast to the previously defined functions, the absolute value of an interval is an interval.

Example 2.1.3 *Let* $x = [-2, 3]$ *and* $y = [1, \pi]$. *Then* $\text{mag}(x) = 3$, $\text{mig}(x) = 0$, $\text{mag}(y) = \pi$, $\text{mig}(y) = 1$, $\text{abs}(x) = [0, 3]$, *and* $\text{abs}(y) = [1, \pi]$.

Finally, we can turn $\mathbb{IR}$ into a metric space (see Definition A.3) by equipping it with the Hausdorff distance:

$$d(\mathbf{a}, \mathbf{b}) = \max\{|\underline{a} - \underline{b}|, |\overline{a} - \overline{b}|\}. \tag{2.1}$$

Note that according to our definitions, it follows that $d(\mathbf{a}, \mathbf{b}) = 0$ if and only if $\mathbf{a} = \mathbf{b}$. Using the metric, we can define the notion of a convergent sequence of intervals:

$$\lim_{k \to \infty} \mathbf{a}_k = \mathbf{a} \quad \Leftrightarrow \quad \lim_{k \to \infty} d(\mathbf{a}_k, \mathbf{a}) = 0$$

$$\Leftrightarrow \quad \left(\lim_{k \to \infty} \underline{a}_k = \underline{a} \right) \wedge \left(\lim_{k \to \infty} \overline{a}_k = \overline{a} \right) \wedge \left(\forall k \quad \underline{a}_k \leq \overline{a}_k \right).$$

Note that the last condition is necessary for the $\Leftarrow$ direction.

2.2 REAL INTERVAL ARITHMETIC

In addition to viewing the elements of $\mathbb{IR}$ as sets, we may consider them as generalized real numbers. As such, it makes sense to attempt to define arithmetic on $\mathbb{IR}$. We have already seen that a copy of $\mathbb{R}$ is represented in $\mathbb{IR}$ as the set of thin intervals. It is therefore desirable to demand that the extended arithmetic should coincide with the normal real arithmetic for thin intervals. The most natural approach is to define binary arithmetic operations on elements of $\mathbb{IR}$ in a set theoretic manner.

DEFINITION 2.1. *If $\star$ is one of the operators $+, -, \times, \div$, we define arithmetic on the elements of $\mathbb{IR}$ by*

$$\mathbf{a} \star \mathbf{b} = \{a \star b \colon a \in \mathbf{a}, b \in \mathbf{b}\},$$

with the exception that $\mathbf{a} \div \mathbf{b}$ is undefined if $0 \in \mathbf{b}$.

From this definition it is not immediately clear that the resulting set always is an interval. As we are working exclusively with closed intervals, however, it turns out that we can describe the resulting set in terms of the endpoints of the operands:

PROPOSITION 2.2.2. *Arithmetic on the elements of $\mathbb{IR}$ is given by*

$$\mathbf{a} + \mathbf{b} = [\underline{a} + \underline{b}, \overline{a} + \overline{b}]$$
$$\mathbf{a} - \mathbf{b} = [\underline{a} - \overline{b}, \overline{a} - \underline{b}]$$
$$\mathbf{a} \times \mathbf{b} = [\min\{\underline{ab}, \underline{a}\overline{b}, \overline{a}\underline{b}, \overline{ab}\}, \max\{\underline{ab}, \underline{a}\overline{b}, \overline{a}\underline{b}, \overline{ab}\}]$$
$$\mathbf{a} \div \mathbf{b} = \mathbf{a} \times [1/\overline{b}, 1/\underline{b}], \quad if \ 0 \notin \mathbf{b}.$$

Proof. The reason why the resulting set is an interval is due to the fact that any *real* operation $+, -, \times, \div$ is continuous in both of its arguments, with the exception of dividing by zero (this is why $\mathbf{a} \div \mathbf{b}$ is undefined[2] if $0 \in \mathbf{b}$). If we fix one of the arguments, the real operations are monotone in the remaining argument. The monotonicity implies that extremal values are attained on the boundary of

---

[2]We can allow for division by zero by extending the underlying set of real numbers to include the concept of infinity. We will address this topic in Section 2.3.

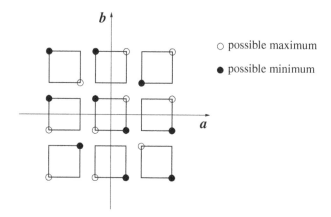

Figure 2.3 A more efficient interval multiplication scheme.

the domains, that is, at the endpoints of the intervals. The proposition can thus be verified by examining a finite number of cases.   □

As a consequence of Proposition 2.2.2, it follows that $\mathbb{IR}$ is an arithmetically closed subset of $\mathcal{P}(\mathbb{R})$— the power set[3] of the real numbers.

From a computer programming point of view, this is good news indeed: using the formulas from Proposition 2.2.2, it is straightforward to implement the datatype `interval` with its associated arithmetic (see Section 2.4). From a practical perspective, the formulas for multiplication and division can be made more efficient. As it stands, a single *interval* multiplication requires four *real* multiplications (as well as several comparisons). This number can be reduced by checking the sign of each endpoint of the two intervals. It is easy to see that interval multiplication can be divided into nine cases, as illustrated in Figure 2.3. Only one case requires four real multiplications; the other cases require just two.

As an example, assume that $0 \leq \underline{a} \leq \overline{a}$ and $\underline{b} \leq 0 \leq \overline{b}$. This situation corresponds to the square on the second row, third column in Figure 2.3. It is clear that the maximal element of $\boldsymbol{a} \times \boldsymbol{b} = \{a \times b : a \in \boldsymbol{a}, b \in \boldsymbol{b}\}$ is given by choosing the largest elements from both $\boldsymbol{a}$ and $\boldsymbol{b}$. By the same token, the minimal element of $\boldsymbol{a} \times \boldsymbol{b}$ is given by choosing the largest element from $\boldsymbol{a}$ and the smallest element from $\boldsymbol{b}$. The resulting interval is thus given by $\boldsymbol{a} \times \boldsymbol{b} = [\overline{a}\underline{b}, \overline{a}\overline{b}]$, which only requires two real multiplications. In a similar fashion, the formula for interval division can be reduced to six simpler cases.

Example 2.2.1 *Using Proposition 2.2.2, we can compute*

$$[-1, 0] + [0, \pi] = [-1, \pi] \quad [-1, -1] \times [2, 5] = [-5, -2]$$

$$[1, 4] - [1, 4] = [-3, 3] \quad [-2, 3] \times [-2, 3] = [-6, 9]$$

$$[\tfrac{1}{2}, 1] - [0, \tfrac{1}{6}] = [\tfrac{1}{3}, 1] \qquad [1, \sqrt{2}] \times [-1, 1] = [-\sqrt{2}, \sqrt{2}]$$

$$[2, 4] - [3, 3] = [-1, 1] \quad [1, 2] \div [-2, -1] = [-2, -\tfrac{1}{2}].$$

---

[3] The power set $\mathcal{P}(\mathbb{S})$ of a set $\mathbb{S}$ is the set of all subsets of $\mathbb{S}$.

It follows from Definition 2.1 (or from Proposition 2.2.2) that addition and multiplication are both associative and commutative: for $a, b, c \in \mathbb{R}$, we have

$$a + (b + c) = (a + b) + c; \qquad a + b = b + a,$$
$$a \times (b \times c) = (a \times b) \times c; \qquad a \times b = b \times a.$$

Also, it is clear that the elements $[0, 0]$ and $[1, 1]$ are the unique neutral elements with respect to addition and multiplication, respectively. Note, however, that in general an element in $\mathbb{R}$ has no additive or multiplicative inverse. For example, we have $[1, 2] - [1, 2] = [-1, 1] \neq [0, 0]$ and $[1, 2] \div [1, 2] = [\frac{1}{2}, 2] \neq [1, 1]$. As a consequence of the arithmetic rules, the distributive law does *not* always hold. As an example,[4] we have

$$[-1, 1]([-1, 0] + [3, 4]) = [-1, 1][2, 4] = [-4, 4],$$

whereas

$$[-1, 1][-1, 0] + [-1, 1][3, 4] = [-1, 1] + [-4, 4] = [-5, 5].$$

This unusual property is important to keep in mind when representing functions as part of an interval calculation. Interval arithmetic satisfies a weaker rule than the distributive law, which we shall refer to as *sub-distributivity*:

$$a(b + c) \subseteq ab + ac. \qquad (2.2)$$

This is a set theoretical property that illustrates one of the fundamental differences between real and interval arithmetic.

Exercise 2.2. *Prove that the space $\mathbb{R}$ can be partially ordered by either relation $\subseteq$ or $\leq$, as described in Section 2.1.*

Exercise 2.3. *Prove that $a(b + c) = ab + ac$ when*

(1) a is thin, or
(2) all elements of b and c have the same sign.

Exercise 2.4. *Given an interval a show that*

(1) $0 \in a - a$, but that in general $a - a \neq [0, 0]$,
(2) $1 \in a \div a$, but that in general $a \div a \neq [1, 1]$. (Assume $0 \notin a$.)

Another key feature of interval arithmetic is that of *inclusion isotonicity*:

THEOREM 2.5. *If $a \subseteq a'$, $b \subseteq b'$, and $\star \in \{+, -, \times, \div\}$, then*

$$a \star b \subseteq a' \star b',$$

where we demand that $0 \notin b'$ for division.

---

[4] Here, and in what follows, we will often suppress the multiplication symbol $\times$.

This is the single most important property of interval arithmetic: it allows us to accurately estimate the range of a large class of functions. This will be explained in a later section. Note that, in particular, Theorem 2.5 holds when $\boldsymbol{a}$ and $\boldsymbol{b}$ are thin intervals, that is, real numbers.

Proof. It is somewhat amazing that this powerful theorem has a classical "one-line" proof: by an immediate application of Definition 2.1, we have

$$\boldsymbol{a} \star \boldsymbol{b} = \{a \star b : a \in \boldsymbol{a}, b \in \boldsymbol{b}\} \subseteq \{a \star b : a \in \boldsymbol{a}', b \in \boldsymbol{b}'\} = \boldsymbol{a}' \star \boldsymbol{b}'.$$

$\square$

2.3 EXTENDED INTERVAL ARITHMETIC

According to Definition 2.1, we cannot divide by an interval containing zero. Nevertheless, if we attempt to reinterpret the spirit of the formula

$$\boldsymbol{a} \div \boldsymbol{b} = \{a \div b : a \in \boldsymbol{a}, b \in \boldsymbol{b}\}$$

as

$$\boldsymbol{a} \div \boldsymbol{b} = \{c \in \mathbb{R} : bc = a, \ a \in \boldsymbol{a}, b \in \boldsymbol{b}\}, \tag{2.3}$$

there might be a way around this slight imperfection. The procedure, however, is quite delicate, and implementing it on a computer raises some subtle questions regarding how we choose to extend the real numbers to include the concept of infinity. Before going into details, let us illustrate the use of (2.3) in a simple setting.

Example 2.3.1 *If $a = [1, 2]$ and $b = [-5, 3]$, then according to (2.3), the quotient $\boldsymbol{c} = \boldsymbol{a} \div \boldsymbol{b}$ is given by*

$$\boldsymbol{c} = \{c \in \mathbb{R} : bc = a, \ a \in [1, 2], b \in [-5, 3]\}.$$

Focusing on the particular value $b = 0$, we want to find all c such that $0 \cdot c \in [1, 2]$. As the equation clearly has no solution, we lose no information by discarding this case. Hence

$$\begin{aligned} \boldsymbol{c} &= \{c \in \mathbb{R} : bc = a, \ a \in [1, 2], b \in [-5, 0) \cup (0, 3]\} \\ &= \{c \in \mathbb{R} : bc = a, \ a \in [1, 2], b \in [-5, 0)\} \cup \\ &\quad \{c \in \mathbb{R} : bc = a, \ a \in [1, 2], b \in (0, 3]\} \\ &= \big([1, 2] \div [-5, 0)\big) \cup \big([1, 2] \div (0, 3]\big). \end{aligned}$$

The first set may be interpreted as the limit

$$\begin{aligned} \boldsymbol{c}^- &= \lim_{\varepsilon \to 0^-} \{c \in \mathbb{R} : bc = a, \ a \in [1, 2], b \in [-5, \varepsilon)\} \\ &= \lim_{\varepsilon \to 0^-} [1, 2] \div [-5, \varepsilon) = \lim_{\varepsilon \to 0^-} (\tfrac{2}{\varepsilon}, -\tfrac{1}{5}) = (-\infty, -\tfrac{1}{5}]. \end{aligned}$$

Similarly, the second set may be interpreted as the limit

$$\begin{aligned} \boldsymbol{c}^+ &= \lim_{\varepsilon \to 0^+} \{c \in \mathbb{R} : bc = a, \ a \in [1, 2], b \in (\varepsilon, 3]\} \\ &= \lim_{\varepsilon \to 0^+} [1, 2] \div (\varepsilon, 3] = \lim_{\varepsilon \to 0^+} [\tfrac{1}{3}, \tfrac{2}{\varepsilon}) = [\tfrac{1}{3}, \infty). \end{aligned}$$

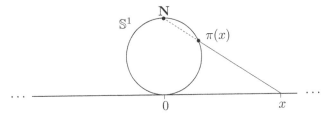

Figure 2.4 Identifying $\mathbb{R}^*$ and $\mathbb{S}^1$ via the projection $\pi : \mathbb{R}^* \to \mathbb{S}^1$.

Combining the two results, we have the answer

$$[1, 2] \div [-5, 3] = (-\infty, -\tfrac{1}{5}] \cup [\tfrac{1}{3}, \infty) = \mathbb{R} \setminus (-\tfrac{1}{5}, \tfrac{1}{3}).$$

This example indicates that we need a notion of infinity in order to perform the extended interval division. There are several ways we can allow for "division by zero"—it all boils down to how we choose to extend the real numbers. From a mathematical point of view, there are more or less elegant extensions. We will acquaint ourselves with three variants, which are appropriately named: *the good, the bad,* and *the ugly.* Naturally, we shall stick to the ugly.

2.3.1 The Good: Projective Extension

The projective extension of the real numbers, usually denoted $\mathbb{R}^*$, is formed by adding the unsigned "point at infinity" ∞ to the real line. This one-point compactification of the real line allows us to identify $\mathbb{R}^*$ with the closed circle $\mathbb{S}^1$ where the north-pole **N** plays the role of infinity (see Figure 2.4).

We can partially extend the arithmetic operations from $\mathbb{R}$ to $\mathbb{R}^*$ in the following manner:

$$-(\infty) = \infty, \qquad\qquad x + \infty = \infty + x = \infty \text{ if } x \neq \infty,$$
$$x \cdot \infty = \infty \cdot x = \infty \text{ if } x \neq 0, \qquad x/\infty = 0 \text{ if } x \neq \infty,$$
$$x/0 = \infty \text{ if } x \neq 0.$$

The expressions $\infty \pm \infty$, ∞/∞, and $0 \cdot \infty$, however, are undefined.

In this setting there is no need for the interpretation (2.3). Instead, repeating the division performed in Example 2.3.1, we immediately have

$$[1, 2] \div [-5, 3] = [1, 2] \div \big([-5, 0) \cup \{0\} \cup (0, 3] \big)$$
$$= \big([1, 2] \div [-5, 0) \big) \cup \big([1, 2] \div 0 \big) \cup \big([1, 2] \div (0, 3] \big)$$
$$= \{ x \in \mathbb{R} : x \leq -\tfrac{1}{5} \} \cup \{\infty\} \cup \{ x \in \mathbb{R} : \tfrac{1}{3} \leq x \}.$$

Note that we no longer can write $(-\infty, -\tfrac{1}{5})$ for $\{ x \in \mathbb{R} : x \leq -\tfrac{1}{5} \}$. This is because, in the projective extension, we have $-\infty = \infty$; that is, there is only one infinity, and it cannot be compared to the finite real numbers with any of the relations $\{<, \leq, >, \geq\}$. As a consequence, we cannot assign the value zero to an expression like $e^{-\infty}$, since in $\mathbb{R}^*$ the equality $e^{-\infty} = e^{\infty}$ must hold. On the other hand, it

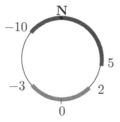

Figure 2.5 The two intervals $[-3, 2]$ and $[5, -10]$ in $\mathbb{R}^*$.

makes perfect sense to write $\tan(\pi/2) = \infty$. The beautiful part of the projective extension is the way we can represent the result of a "division by zero." So far, the outcome of performing $[1, 2] \div [-5, 3]$ appears to be a rather messy expression. Nevertheless, using the topology of the circle, we may adopt a shorthand notation for *extended intervals*:

$$[\tfrac{1}{3}, -\tfrac{1}{5}] = \{x \in \mathbb{R}: x \le -\tfrac{1}{5}\} \cup \{\infty\} \cup \{x \in \mathbb{R}: \tfrac{1}{3} \le x\}.$$

To motivate this notation, we refer to Figure 2.5, which shows two intervals in $\mathbb{R}^*$. The interval containing zero is simply the set $\{x \in \mathbb{R}^*: -3 \le x \text{ and } x \le 2\}$, which we denote $[-3, 2]$ as usual. The second interval, however, is different. It represents the set $\{x \in \mathbb{R}^*: x \le -10 \text{ or } 5 \le x \text{ or } x = \infty\}$, which we can write as $[5, -10]$. This should be interpreted on the circle as moving from the left endpoint, 5, counterclockwise to the right endpoint, -10, just as with normal intervals. Unfortunately, there is no suitable way to represent the extended real line $\mathbb{R}^*$ in this manner.

Motivated by the preceding discussion, we define the set $\mathbb{R}^*$ of projectively extended intervals:

$$\mathbb{R}^* = \{[\underline{a}, \overline{a}]: \quad \underline{a}, \overline{a} \in \mathbb{R}^*\},$$

where the case $\underline{a} > \overline{a}$ is interpreted as an extended interval.

2.3.2 The Bad: Affine Extension

The affine extension of the real numbers, usually denoted $\overline{\mathbb{R}}$, is formed by adding the two signed infinities, $-\infty$ and $+\infty$, to the real line. This two-point compactification of the real line allows us to write $\overline{\mathbb{R}}$ as the closed interval $[-\infty, +\infty]$. The arithmetic operations can be partially extended to $\overline{\mathbb{R}}$ in the following manner:

$$-(+\infty) = -\infty \text{ and } -(-\infty) = +\infty, \qquad x + (+\infty) = +\infty \text{ if } x \ne -\infty,$$
$$x + (-\infty) = -\infty \text{ if } x \ne +\infty, \qquad x \cdot (\pm\infty) = \pm\infty \text{ if } x > 0,$$
$$x \cdot (\pm\infty) = \mp\infty \text{ if } x < 0, \qquad x/(\pm\infty) = 0 \text{ if } x \ne \pm\infty.$$

The expressions $+\infty + (-\infty)$, $-\infty + (+\infty)$, and $x/0$, however, are undefined. In contrast to the projective extension, the affine infinities can be compared in size: $-\infty < x < +\infty$ if $x \ne \pm\infty$ and $-\infty < +\infty$. Furthermore, the affine extension has the appealing property that it makes perfect sense to write statements like

$e^{-\infty} = 0$, $e^{+\infty} = +\infty$, $\ln 0 = -\infty$, and $\ln(+\infty) = +\infty$. This property makes the affine extension the preferred choice among analysts.

It is now straightforward to define the set $\overline{\mathbb{R}}$ of affinely extended intervals:

$$\overline{\mathbb{IR}} = \{[\underline{a}, \overline{a}] : \underline{a} \le \overline{a}; \quad \underline{a}, \overline{a} \in \overline{\mathbb{R}}\}.$$

Thus, apart from the elements of $\mathbb{IR}$, intervals on the form $[-\infty, x]$, $[x, +\infty]$, and $[-\infty, +\infty]$ are also valid elements of $\overline{\mathbb{IR}}$.

Since we cannot divide by zero in $\overline{\mathbb{R}}$, repeating the division performed in Example 2.3.1 requires the interpretation (2.3) and produces

$$[1, 2] \div [-5, 3] = [-\infty, -\tfrac{1}{5}] \cup [\tfrac{1}{3}, \infty].$$

As with the projective extension, we could simply introduce the notion of extended intervals (thus removing the demand $\underline{a} \le \overline{a}$ for intervals), now with the meaning

$$[\tfrac{1}{3}, -\tfrac{1}{5}] = [-\infty, -\tfrac{1}{5}] \cup [\tfrac{1}{3}, \infty]. \tag{2.4}$$

Alternatively, we could accept the fact that some interval operations may produce a union of intervals. This line of action, however, leads to some tricky implementation issues. A simpler way of resolving the whole issue would be to return the entire line $[-\infty, +\infty]$ when dividing by zero, possibly with the exception that $\mathbf{a}/0 = \emptyset$ if $0 \notin \mathbf{a}$. As this approach leads to an unnecessary loss of information, it is therefore not very widespread. We will use the definition

$$\overline{\mathbb{IR}} = \{[\underline{a}, \overline{a}] : \underline{a}, \overline{a} \in \overline{\mathbb{R}}\}$$

where the case $\underline{a} \ge \overline{a}$ corresponds to an extended interval of type (2.4).

2.3.3 The Ugly: Signed Zero

While all elements of $\mathbb{R}^*$ have unique reciprocals, this is not the case for all members of $\overline{\mathbb{R}}$. Indeed, in $\overline{\mathbb{R}}$ we have $1/(-\infty) = 1/(+\infty) = 0$, whereas $1/0$ is undefined. In an attempt to resolve this problem, it is possible to equip $\overline{\mathbb{R}}$ with *signed zeros*, satisfying $x/(+0) = \text{sign}(x) \cdot (+\infty)$ and $x/(-0) = \text{sign}(x) \cdot (-\infty)$ for $x \ne \pm 0$. In effect, this gives both infinites and both signed zeros unique reciprocals, just like all other elements of $\overline{\mathbb{R}}$:

$$1/(+\infty) = +0, \quad 1/(+0) = +\infty, \quad 1/(-\infty) = -0, \quad 1/(-0) = -\infty.$$

As an illustration, we have $1 \div [+0, 2] = [\tfrac{1}{2}, +\infty]$, whereas $1 \div [-0, 2] = [-\infty, +\infty]$. Unfortunately, there is no natural way of propagating the sign of the zero under addition and subtraction: what sign should $(+0) + (-0)$ or even $x - x$ have?

The IEEE standard incorporates signed infinities as well as signed zeros. The signs appear naturally within the actual sign-exponent-mantissa encoding of the floating point numbers. Even though -0 and $+0$ are distinct values, they both compare as equal and are only distinguishable by comparing their sign bits. Regarding addition and subtraction, the standard [IE85] states

> When the sum of two operands with opposite signs (or the difference of two operands with equal signs) is exactly zero, the sign of that sum

(or difference) shall be $+$ in all rounding modes except round toward $-\infty$, in which mode that sign shall be $-$. However, $x + x = x - (-x)$ retains the same sign as x even when x is zero.

Thus, on any computer compliant with the IEEE standard, we have $(+0) + (-0) = x - x = +0$, unless we are rounding with $\triangledown$, answering the question posed above.

The signed zeros were not introduced for their mathematical elegance: their presence is due to computer manufacturers' desire to reduce the number of fatal floating point errors. Instead of having to abort a computation that happens to perform a division by zero, it is much more desirable to produce a well-defined result of the division. Without the signed zero, this is simply not possible.

2.3.4 The Extended Interval Division

In light of the previous discussion, the extended interval division is defined over the space $\overline{\mathbb{IR}}$ (equipped with signed zeros), where we allow for extended intervals of the form (2.4). Following [Ra96], we define division over $\overline{\mathbb{IR}}$ as follows

$$a \div b = \begin{cases} a \times [1/\overline{b}, 1/\underline{b}] & \text{if } 0 \notin b, \\ [-\infty, +\infty] & \text{if } 0 \in a \text{ and } 0 \in b, \\ [\overline{a}/\underline{b}, +\infty] & \text{if } \overline{a} < 0 \text{ and } \underline{b} < \overline{b} = 0, \\ [\overline{a}/\underline{b}, \overline{a}/\overline{b}] & \text{if } \overline{a} < 0 \text{ and } \underline{b} < 0 < \overline{b}, \\ [-\infty, \overline{a}/\overline{b}] & \text{if } \overline{a} < 0 \text{ and } 0 = \underline{b} < \overline{b}, \\ [-\infty, \underline{a}/\underline{b}] & \text{if } 0 < \underline{a} \text{ and } \underline{b} < \overline{b} = 0, \\ [\underline{a}/\overline{b}, \underline{a}/\underline{b}] & \text{if } 0 < \underline{a} \text{ and } \underline{b} < 0 < \overline{b}, \\ [\underline{a}/\overline{b}, +\infty] & \text{if } 0 < \underline{a} \text{ and } 0 = \underline{b} < \overline{b}, \\ \emptyset & \text{if } 0 \notin a \text{ and } b = [0, 0]. \end{cases} \quad (2.5)$$

Case 1 deals with non-zero divisors, although it now incorporates quotients such as $[6, 8] \div [2, +\infty] = [+0, 4]$. Cases 4 and 7 yield extended intervals, that is, these particular results actually consist of a union of two infinite intervals. In [Ra96], it is proved that the division defined by (2.5) is inclusion isotonic, that is, if $a \subseteq a'$, and $b \subseteq b'$, then $a \div b \subseteq a' \div b'$, which generalizes Theorem 2.5.

It is worth pointing out that although signed zeros are not explicitly present in (2.5), their properties are mimicked in the formulas. As an example, let us consider case 5, where the condition is stated as "if $\overline{a} < 0$ and $0 = \underline{b} < \overline{b}$." For all practical purposes, this can be interpreted as "if $\overline{a} < 0$ and $+0 = \underline{b} < \overline{b}$."

There are two main advantages of having access to an extended interval division in a computing environment. First, all run-time errors of the type *division by zero* are immediately avoided. Usually, an error of this type will cause a program to crash, unless some serious error-handling capabilities have been provided by the programmer. Second, it is actually desirable, from a mathematical point of view, to be able to perform extended division. Later on, we will see a striking example of this when we study the interval Newton method.

2.3.5 Containment Sets

In many situations, not even an extended interval arithmetic will suffice. In a set-valued environment there are many more ways in which a seemingly innocent calculation can result in a run-time error. As an example, consider the function $f(x) = \sqrt{x - x^2}$ evaluated over the domain $x = [0, 1]$. It is clear that the exact range[5] of f is well-defined on this domain: $R(f; [0, 1]) = [0, 1/\sqrt{2}]$. The interval evaluation, however, is problematic:

$$F([0, 1]) = \sqrt{[0, 1] - [0, 1]^2} = \sqrt{[0, 1] - [0, 1]} = \sqrt{[-1, 1]}.$$

There are (at least) three ways to handle this situation: (1) we can abort the computation; (2) we can give a complex-valued result; or (3) we can restrict the domain before taking the square root.

Resorting to (3) is usually referred to as *loose evaluation*. More precisely, for any domain x, we define

$$\sqrt{x} = \sqrt{x \cap [0, +\infty]}.$$

For the example considered above, this yields:

$$F([0, 1]) = \sqrt{[-1, 1]} = \sqrt{[-1, 1] \cap [0, +\infty]} = \sqrt{[0, 1]} = [0, 1].$$

The idea is to disregard arguments that do not belong to the function's natural domain. Note that although we do not obtain the exact range of the function, we still have an enclosure: $R(f; [0, 1]) \subseteq F([0, 1])$. This is paramount to interval computations and will be further addressed in Section 3.1.

The idea of containment sets (csets) takes this idea further. The aim is to create a completely *exception free* system, that is, one where all operations are always well-defined. Given an elementary, real-valued function $f: D_f \to \mathbb{R}$, where D_f is the largest domain on which f is well-defined, we introduce the cset extension $f^*: \mathcal{P}\mathbb{R} \to \mathcal{P}\mathbb{R}$ via

$$f^*(S) = R(f; S \cap D_f) \cup \{\lim_{\zeta \to \zeta^*} f(\zeta): \zeta \in D_f, \zeta^* \in S \setminus D_f\}. \qquad (2.6)$$

Here $\overline{\mathbb{R}}$ is the set of affinely extended reals $\overline{\mathbb{R}} = \{-\infty\} \cup \mathbb{R} \cup \{+\infty\}$, and $\mathcal{P}\overline{\mathbb{R}}$ is the set of all subsets of $\overline{\mathbb{R}}$ – the power set of $\overline{\mathbb{R}}$. The definition (2.6) says that we consider the cset extension to be made up of two parts: one part for all points that belong to f's natural domain D_f (the loose evaluation), and one part for points that are located just outside D_f.

Returning to the example function $f(x) = \sqrt{x - x^2}$, we have $f^\star([0.1]) = R(f; [0, 1]) = [0, 1/\sqrt{2}]$, seeing that no arguments are outside the natural domain of f. In a realistic implementation (where only standard functions are optimally implemented), we would probably overestimate the range of $x - x^2$ with the interval $[-1.1]$. This would transform the original problem to that involving $g(x) = \sqrt{x}$

---

[5] The range is defined as $R(f; S) = \{f(x): x \in S\}$.

with $D_g = [0, +\infty)$ and $S = [-1, 1]$. Using (2.6), we get

$$g^*([-1, 1]) = R(g; [-1, 1] \cap [0, +\infty)) \cup \{\lim_{\zeta \to \zeta^*} g(\zeta):$$

$$\zeta \in [0, +\infty), \zeta^* \in [-1, 1] \setminus [0, +\infty)\}$$

$$= R(g; [0, 1]) \cup \{\lim_{\zeta \to \zeta^*} g(\zeta): \zeta \in [0, +\infty), \zeta^* \in [-1, 0)\}$$

$$= g([0, 1]) \cup \emptyset = \sqrt{[0, 1]} = [0, 1],$$

which corresponds to a pure domain restriction, that is, loose evaluation.

A more interesting example is given by taking $h(x) = 1/x$, which has the natural domain $D_h = (-\infty, 0) \cup (0, +\infty)$. Again, by using (2.6), we get for $S = \{0\}$

$$h^*(0) = R(h; \{0\} \cap \{(-\infty, 0) \cup (0, +\infty)\}) \cup \{\lim_{\zeta \to \zeta^*} h(\zeta):$$

$$\zeta \in (-\infty, 0) \cup (0, +\infty), \zeta^* \in \{0\} \setminus \{(-\infty, 0) \cup (0, +\infty)\}\}$$

$$= R(h; \emptyset) \cup \{\lim_{\zeta \to 0} h(\zeta): \zeta \in (-\infty, 0) \cup (0, +\infty)\}$$

$$= \emptyset \cup \{\lim_{\zeta \to 0^-} h(\zeta)\} \cup \{\lim_{\zeta \to 0^+} h(\zeta)\} = \{-\infty\} \cup \{+\infty\}.$$

Compare this result to the last line of (2.5), which defines the outcome as the empty set. We can redo Example 2.3.1 by taking $S = [-3, 5]$:

$$h^*([-3, 5]) = R(h; [-3, 5] \cap \{(-\infty, 0) \cup (0, +\infty)\}) \cup \{\lim_{\zeta \to \zeta^*} f(\zeta):$$

$$\zeta \in (-\infty, 0) \cup (0, +\infty), \zeta^* \in [-3, 5] \setminus \{(-\infty, 0) \cup (0, +\infty)\}\}$$

$$= R(h; [-3, 0) \cup (0, 5]) \cup \{\lim_{\zeta \to 0} h(\zeta): \zeta \in (-\infty, 0) \cup (0, +\infty)\}$$

$$= (-\infty, -\tfrac{1}{3}] \cup [\tfrac{1}{5}, +\infty) \cup \{\lim_{\zeta \to 0^-} h(\zeta)\} \cup \{\lim_{\zeta \to 0^+} h(\zeta)\}$$

$$= (-\infty, -\tfrac{1}{3}] \cup [\tfrac{1}{5}, +\infty) \cup \{-\infty\} \cup \{+\infty\}$$

$$= [-\infty, -\tfrac{1}{3}] \cup [\tfrac{1}{5}, +\infty].$$

This is exactly the same result that was obtained in Section 2.3.2.

Now, for an interval version of csets, we can return the interval hull of the result. Using the examples above, we get $F^*([0, 1]) = G^*([-1, 1]) = [0, 1]$, $H^*([0, 0]) = H^*([-3, 5]) = [-\infty, +\infty]$. A really nice feature of this definition is that the interval cset arithmetic is exception free: that is, there are no undefined operations. As a consequence, we always get a true enclosure of the range:

$$R(f; x) \subseteq F^*(x).$$

For a clear exposition of theoretical as well as practical points of containment sets, see [PC06]. SUN Microsystems' C++ and Fortran compilers support interval arithmetic with cset functionality (see [SUN7]). The free C++ library FILIB++ also has this functionality as an option (see [LT06]).

Exercise 2.6. *A special case of Brouwer's fixed point theorem ([Br10]) states that any continuous function $f : [-1, 1] \to [-1, 1]$ has a fixed point x^* in $[-1, 1]$, that is, $f(x^*) = x^*$. If $f(x) = \sqrt{x} - 1$, and F^* is the cset extension of f, then we have*

$R(f; [-1, 1]) \subseteq F^*([-1, 1]) = [-1, 0]$ *(show this). This means that* f *indeed maps* $[-1, 1]$ *into itself and by Brouwer's theorem* f *should have a fixed point in* $[-1, 1]$, *which it does not. Explain.*

2.4 FLOATING POINT INTERVAL ARITHMETIC

When implementing interval arithmetic on a computer, we no longer work over the space $\mathbb{R}$ but rather $\mathbb{F}$—the floating point numbers of the computer. This is a finite set, as is $\mathbb{IF}$—the set of all intervals whose endpoints belong to $\mathbb{F}$:

$$\mathbb{IF} = \{[\underline{a}, \overline{a}] : \underline{a} \le \overline{a}; \quad \underline{a}, \overline{a} \in \mathbb{F}\}.$$

As discussed earlier, $\mathbb{F}$ is not arithmetically closed. Thus, when performing arithmetic on intervals in $\mathbb{F}$ we must round the resulting interval *outward* to guarantee inclusion of the true result. By this, we mean that the lower bound is rounded down, and the upper bound is rounded up. For $\boldsymbol{a}, \boldsymbol{b} \in \mathbb{IF}$, we define

$$\boldsymbol{a} + \boldsymbol{b} = [\triangledown(\underline{a} + \underline{b}), \triangle(\overline{a} + \overline{b})]$$

$$\boldsymbol{a} - \boldsymbol{b} = [\triangledown(\underline{a} - \overline{b}), \triangle(\overline{a} - \underline{b})]$$

$$\boldsymbol{a} \times \boldsymbol{b} = [\min\{\triangledown(\underline{a}\underline{b}), \triangledown(\underline{a}\overline{b}), \triangledown(\overline{a}\underline{b}), \triangledown(\overline{a}\overline{b})\},$$
$$\max\{\triangle(\underline{a}\underline{b}), \triangle(\underline{a}\overline{b}), \triangle(\overline{a}\underline{b}), \triangle(\overline{a}\overline{b})\}]$$

$$\boldsymbol{a} \div \boldsymbol{b} = [\min\{\triangledown(\underline{a}/\underline{b}), \triangledown(\underline{a}/\overline{b}), \triangledown(\overline{a}/\underline{b}), \triangledown(\overline{a}/\overline{b})\},$$
$$\max\{\triangle(\underline{a}/\underline{b}), \triangle(\underline{a}/\overline{b}), \triangle(\overline{a}/\underline{b}), \triangle(\overline{a}/\overline{b})\}], \quad \text{if } 0 \notin \boldsymbol{b}.$$

Recall that $\triangledown(x)$ and $\triangle(x)$ were defined in Section 1.3.2. The resulting type of arithmetic is called interval arithmetic with *directed rounding*. As we shall see, this is easily implemented on a computer that supports the directed roundings.

With regards to efficiency, a single $\mathbb{IF}$-multiplication requires eight $\mathbb{F}$-multiplications: four products must be computed under two different rounding modes. As before, it is customary to break the formula for multiplication into nine cases (depending on the signs of the operands' endpoints). Out of these nine cases, only one will involve four $\mathbb{F}$-multiplications; the remaining eight will need just two. In a similar manner, the (non-extended) $\mathbb{IF}$-division can be split into six cases.

Exercise 2.7. *Derive the optimal formulas for division in* $\mathbb{IF}$, *assuming that a floating point comparison is much faster than a floating point division.*

Extending the floating point interval arithmetic via (2.5) is straightforward and yields the set

$$\overline{\mathbb{IF}} = \{[\underline{a}, \overline{a}] : \underline{a}, \overline{a} \in \overline{\mathbb{F}}\},$$

where the case $\underline{a} \ge \overline{a}$ corresponds to an extended interval of type (2.4).

At the time of writing, a standard (IEEE p1788) for interval arithmetic is in the process of being drafted (see http://grouper.ieee.org/groups/1788/).

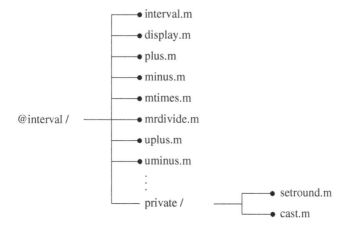

Figure 2.6 The hierarchy of the MATLAB interval class.

2.4.1 A **MATLAB** Implementation of Interval Arithmetic

To illustrate how easy it is to get started, we present a simple MATLAB implementation of (non-extended) interval arithmetic with directed rounding.

As with most modern programming languages, MATLAB uses *classes* to define new data types and *methods* to define the functionality of a user-defined class. A new class can be added to the MATLAB environment by specifying a structure that provides data storage for the object and creating a class directory containing m-files[6] that operate on the object. These m-files contain the methods for the class. MATLAB is somewhat peculiar in that it demands a certain file hierarchy associated with each class. In Figure 2.6 we illustrate a simple setup for our interval class. We will explain the purpose of the different m-files as we go along.

As we want to build an interval class, we begin by creating a directory called @interval where all m-files associated with the interval class will reside. Having done this, we create the m-file interval.m, in which we define what is meant by an interval. It is natural to implement an interval as a class consisting of two numbers—the endpoints of the interval:

```
01 function iv = interval(lo, hi)
02 % A naive interval class constructor.
03 if nargin == 1
04     hi = lo;
05 elseif ( hi < lo )
06     error('The endpoints do not define an interval.');
07 end
08 iv.lo = lo; iv.hi = hi;
09 iv = class(iv,'interval');
```

By including lines 03 and 04, we allow the constuctor to automatically convert a single number x to a thin interval $[x, x]$. Note that, as opposed to most programming languages, MATLAB only supports the double format, which means that no explicit type declarations have to (or can) be made.

---

[6]An m-file is simply a text file filename.m containing a sequence of MATLAB statements to be executed. The file extension of .m makes this a MATLAB m-file.

Table 2.1 Overloaded MATLAB arithmetic methods

Operation	m-file	New description
a + b	plus(a,b)	Interval addition
a - b	minus(a,b)	Interval subtraction
a * b	mtimes(a,b)	Interval multiplication
a / b	rmdiv(a,b)	Interval division
+a	uplus(a)	Unary plus
-a	uminus(a)	Unary minus

We must also inform MATLAB how to display interval objects. This is achieved via the m-file display.m:

```
01 function display(iv)
02 % A simple output formatter for the interval class.
03 disp([inputname(1), ' = ']);
04 fprintf('  [%17.17f, %17.17f]\n', iv.lo, iv.hi);
```

We can now input/output intervals within the MATLAB environment:

```
>> a = interval(3, 4), b = interval(2, 5), c = interval(1)
a =
  [3.00000000000000000, 4.00000000000000000]
b =
  [2.00000000000000000, 5.00000000000000000]
c =
  [1.00000000000000000, 1.00000000000000000]
```

When creating user-defined classes, it is often desirable to change the behavior of the MATLAB operators and functions[7] for cases when the arguments are user-defined classes. This can be accomplished by *overloading* the relevant functions. Overloading enables a function to handle different types and numbers of input arguments and perform whatever operation is appropriate for the situation at hand.

Each native MATLAB operator has an associated function name (e.g., the + operator has an associated plus.m function). Any such operator can be overloaded by creating an m-file with the appropriate name in the class directory. In Table 2.1, we list the operators we intend to overload in our interval class.

Another feature of MATLAB becomes evident: the MATLAB-engine regards all numeric elements as matrices, even if they are single numbers. Indeed, a single number can be viewed as a 1×1 matrix.

Let us begin by writing a function that returns the sum of two intervals:

```
01 function result = plus(a, b)
02 % Overloading the '+' operator for intervals.
03 [a, b] = cast(a, b);
04 setround(-inf);
05 lo = a.lo + b.lo;
06 setround(+inf);
```

---

[7]In what follows, we will not distinguish between *functions* and *methods*.

```
07 hi = a.hi + b.hi;
08 setround(0.5);
09 result = interval(lo, hi);
```

Let us examine this small piece of code. First, the function cast, appearing on line 03, makes sure that the inputs a and b are intervals. If one of them is not an interval, it is converted to an interval by a call to the interval constructor (see the listing below).

```
01 function [a, b] = cast(a, b)
02 % Casts non-intervals to intervals.
03 if ~isa(a, 'interval')
04     a = interval(a);
05 end
06 if ~isa(b, 'interval')
07     b = interval(b);
08 end
```

Casting[8] allows for expressions like $[1, 2] + 3$, which is converted to $[1, 2] + [3, 3]$ and evaluated as $[4, 5]$.

Second, the function setround, appearing on lines 04, 06, and 08 of the file plus.m, instructs the MATLAB-engine to switch the rounding direction before performing an arithmetic operation. This function is implemented[9] in the auxiliary file setround.m, presented below:

```
01 function setround(rnd)
02 % A switch for changing rounding mode. The arguments
03 % {+inf, -inf, 0.5, 0} correspond to the roundings
04 % {upward, downward, to nearest, to zero}, respectively.
05 system_dependent('setround',rnd);
```

We consider both functions cast and setround to be intrinsic to the interval class. By placing their m-files in the private subdirectory, these functions are hidden from non-interval classes.

Carrying on, it is straightforward to write functions that overload the remaining arithmetic operations $-$, $\times$, and $\div$. Below, we present a MATLAB listing of the division algorithm:

```
01 function result = mrdivide(a, b)
02 % A non-optimal interval division algorithm.
03 [a, b] = cast(a, b);
04 if ( (b.lo <= 0.0) & (0.0 <= b.hi) )
05     error('Denominator straddles zero.');
06 else
07     setround(-inf);
08     lo = min(a.lo/b.lo, a.lo/b.hi, a.hi/b.lo, a.hi/b.hi);
09     setround(+inf);
10     hi = min(a.lo/b.lo, a.lo/b.hi, a.hi/b.lo, a.hi/b.hi);
11     setround(0.5);
12     result = interval(lo, hi);
13 end
```

---

[8]In the programming language C++, casting is implicitly performed at the compilation stage. This simplifies the actual programming but can also produce hard-to-find bugs.

[9]Unfortunately, this feature is not available on SPARC platforms. Instead, a small C program must be pre-compiled into a so-called mex-file (see the code in Section 2.4.2).

Performing some simple interval calculations, we have:

```
>> a+b, a-b, a*b, a/b
ans =
   [5.0000000000000000, 9.00000000000000000]
ans =
   [-2.00000000000000000, 2.00000000000000000]
ans =
   [6.00000000000000000, 20.00000000000000000]
ans =
   [0.5999999999999998, 2.00000000000000000]
```

The outward rounding is apparent in the left endpoint of the last result. All other endpoints were computed exactly. We should point out that our interval constructor is still very rudimentary and does not handle user input adequately. As an example, suppose we would like to generate the smallest interval containing $1/10$. As a first attempt, we may try something like

```
>> interval(1/10)
ans =
   [0.1000000000000001, 0.10000000000000001]
```

which is *not* what we wanted. The problem here is that the quotient $1/10$ is *first* rounded to a single floating point number, which is *then* converted to a thin interval. Since $1/10$ has no exact representation in the floating point format, we obtain an interval that does not contain $1/10$. A way to work around this is to declare either the nominator or denominator as an interval. Since integers have exact representations, no rounding takes place at this stage. It is only when the division takes place that the directed rounding kicks in, producing a non-thin interval straddling $1/10$:

```
>> interval(1)/10
ans =
   [0.09999999999999999, 0.10000000000000001]
```

More sophisticated interval libraries provide a means for entering strings of numbers,[10] such as

```
>> interval('1/10')
ans =
   [0.09999999999999999, 0.10000000000000001]
```

Nevertheless, this has a cost in programming effort that we are not willing to pay at the moment.

Continuing our calculations, we can now illustrate the sub-distributive property of interval arithmetic:

```
>> c = interval(0.25, 0.50)
c =
   [0.2500000000000000, 0.50000000000000000]
>> a*(b+c), a*b+a*c
ans =
   [5.2500000000000000, 22.00000000000000000]
ans =
   [5.0000000000000000, 22.00000000000000000]
```

---

[10]The MATLAB package IntLab has this functionality, as does the C++ toolbox CXSC (see [Ru99] and [CXSC], respectively).

Table 2.2 Overloaded MATLAB set-relation methods

Operation	m-file	New description
a == b	eq(a,b)	Equal to
a ~= b	ne(a,b)	Not equal to
a <= b	le(a,b)	Subset of
a < b	lt(a,b)	Proper subset of
a & b	and(a,b)	Intersection
a | b	or(a,b)	Interval hull

Notice the differing lower endpoints; clearly the expression ab + ac produces a wider result than a(b + c). Since all computations in this example are exact, the outward rounding does not affect the result.

Exercise 2.8. *Modify the appropriate m-files so they perform multiplication and division by checking the signs of the operands' endpoints.*

Exercise 2.9. *Add some interval functions (e.g.,* sin(x) *and* pow(x,n)*) to the* interval *class. Note that this requires some knowledge of how accurate the corresponding real-valued functions are in the underlying programming environment.*

Exercise 2.10. *Do you know any other programming language that supports operator overloading? If so, try to implement a rudimentary interval arithmetic library whose syntax permits expressions like* x + y *and* z = sin(1 + pow(x,2))*, where* x, y, *and* z *are of type* interval*.*

Now that we have all arithmetic operations in place, let us consider the built-in relational operators provided by MATLAB. Some of these are listed in Table 2.2.

When overloading these methods, we will give them new, interval-based meanings. Let us begin with the simplest of them all: the *equality* relation. Two intervals are *equal* exactly when their endpoints agree. Analogously, two intervals are *not equal* if at least one of their endpoints differs. Both functions can be implemented in a few lines.

```
01 function result = eq(a, b)
02 % The '(e)qual' operator '=='.
03 [a, b] = cast(a, b);
04 result = ( (a.lo == b.lo) & (a.hi == b.hi) );
```

```
01 function result = ne(a, b)
02 % The '(n)ot (e)qual' operator '~='.
03 [a, b] = cast(a, b);
04 result = ( (a.lo ~= b.lo) | (a.hi ~= b.hi) );
```

Turning to the order-relations *less or equal* and *less than*, we will interpret them as the set-relations *inclusion* $\subseteq$ and *proper inclusion* $\overset{\circ}{\subset}$, respectively.

```
01 function result = le(a, b)
02 % The '(l)ess or (e)qual' operator '<='. Means 'a inside b'.
03 [a, b] = cast(a, b);
04 result = ( (b.lo <= a.lo) & ( a.hi <= b.hi) );
```

```
01 function result = lt(a, b)
02 % The '(l)ess (t)han' operator '<'. Means 'a inside int(b)',
03 [a, b] = cast(a, b);
04 result = ( (b.lo < a.lo) & ( a.hi < b.hi) );
```

The four functions we have defined so far are all *boolean*, that is, their return-values come from the set {true, false}. In MATLAB (and most other programming languages), these alternatives are coded as '1' and '0', respectively.

```
>> a = interval(1, 10); b = interval(-2, 3);
c = interval(3, 5);
>> [a==b, a==c, a~=b, a~=c, a<=b, b<=a, a<c, c<a]
ans =
      0     0     1     1     0     0     0     1
```

Finally, we will implement the logical operations *and* and *or*, but we will redefine them as the set-operations *intersection* ∩ and *hull* ⊔, respectively. One complication here is that two intervals may have an empty intersection. Seeing that our simple interval constructor does not accommodate empty intervals, we will return the MATLAB version of the empty set, accompanied by a warning[11] whenever this situation occurs.

```
01 function result = and(a, b)
02 % The 'and' operator '&'. Means 'a intersected with b'.
03 [a, b] = cast(a, b);
04 if ( (a.hi < b.lo) | (b.hi < a.lo) )
05     warning('The intervals do not intersect.');
06     result = [];
07 else
08     result = interval(max(a.lo, b.lo), min(a.hi, b.hi));
09 end
```

Since we have not defined the interval methods to operate on empty sets, it is vital that we have a means for detecting an empty set. Fortunately, MATLAB has a built-in function isempty that can reveal whether the outcome of an interval intersection is an empty set or not.

```
>> a=interval(1,3); b=interval(4,5); c=interval(2,5);
>> aANDb = a & b; aANDc = a & c;
Warning: The intervals do not intersect.
> In /matlab/@interval/and.m at line 5
aANDb =
     []
aANDc =
  [2.00000000000000000, 3.00000000000000000]
>> [isempty(aANDb), isempty(aANDc)]
ans =
      1     0
```

---

[11] It may be desirable to comment out the warning in order to minimize unnecessary output.

The hull of two intervals is always well-defined and thus straightforward to implement.

```
01 function result = or(a, b)
02 % The 'or' operator '|'. Means 'hull of a and b'.
03 [a, b] = cast(a, b);
04 result = interval(min(a.lo, b.lo), max(a.hi, b.hi));
```

```
>> aORb = a | b, aORc = a | c
aORb =
   [1.00000000000000000, 5.00000000000000000]
aORc =
   [1.00000000000000000, 5.00000000000000000]
```

Exercise 2.11. *Make the necessary modifications to the m-files* interval.m *and* display.m *to accommodate input/output of empty intervals. As a first step you must find a good representation for the empty interval.*

Exercise 2.12. *How would you modify the remaining m-files to fully incorporate empty intervals?*

Exercise 2.13. *Implement a new class* xinterval *for extended intervals. Try to extend all associated interval methods described in this section. (Warning: this takes some effort.)*

Exercise 2.14. *An interval can also be represented by the two numbers $< m, r >$, where m is the midpoint, and r is the radius of the interval. Derive the interval-arithmetic rules using only these two quantities. From a numerical point of view, what are the merits and/or drawbacks of this (*midrad*) representation compared to the endpoint (*infsup*) representation?*

Exercise 2.15. *Download and install* IntLab *(see [Ru99]) on your system. Try some of the demonstration programs to explore the many features of the package.*

2.4.2 Changing Rounding Modes

In Listing 2.1 we present a header file that allows the rounding direction to be changed from within a C/C++ program. A word of caution, though: it appears that some 64-bit architectures (AMD64, MIPS, PPC, HP/PA, S/390, SPARC, Alpha) do not work properly with respect to the header fenv.h. Always run a few checks to make sure that the rounding instructions are in action. If your platform of choice does not round correctly in 64-bit mode, it is possible to compile to 32-bit code, which should produce proper machine code.

As mentioned in Section 2.4.1, all platforms do not support the MATLAB command system_dependent. In order to switch rounding mode one can instead compile a small C program (see Appendix B.2.1) containing the switching instructions into a module that MATLAB can use. This is done via MATLAB's compiler mex.

Listing 2.1.   A header for switching rounding mode within C/C++ codes

```
1  /*    File: round.h
2
3       A header file that defines the directed
4       rounding modes for Linux and Sparc.
5  */
6
7  #if defined(__linux)
8  #include <fenv.h>
9  #define ROUND_DOWN   FE_DOWNWARD
10 #define ROUND_UP     FE_UPWARD
11 #define ROUND_NEAR   FE_TONEAREST
12 #define setRound     fesetround
13 #endif
14
15 #if defined(__sparc)
16 #include <ieeefp.h>
17 #define ROUND_DOWN   FP_RM
18 #define ROUND_UP     FP_RP
19 #define ROUND_NEAR   FP_RN
20 #define setRound     fpsetround
21 #endif
22
23 void setRoundDown() { setRound(ROUND_DOWN); }
24 void setRoundUp  () { setRound(ROUND_UP);   }
25 void setRoundNear() { setRound(ROUND_NEAR); }
```

Chapter Three

Interval Analysis

3.1 INTERVAL FUNCTIONS

One of the main reasons for studying interval arithmetic is that we want a simple way of enclosing the *range* of a real-valued function.

DEFINITION 3.1. *Let $D \subseteq \mathbb{R}$, and consider a function $f : D \to \mathbb{R}$. We define the range of f over D to be the set*

$$R(f; D) = \{f(x): x \in D\}.$$

Except for the most trivial cases, mathematics provides few tools to describe the range of a given function f over a specific domain D. Indeed, today there is an entire branch of mathematics and computer science—optimization theory— devoted to "simply" finding the smallest element of the set $R(f; D)$. We shall see that interval arithmetic provides a helping hand in this matter.

As a first step, we begin by attempting to extend the real functions to *interval functions*. By this, we mean functions that take and return intervals rather than real numbers. We already have the theory to extend rational functions, that is, functions on the form $f(x) = p(x)/q(x)$, where p and q are polynomials. Simply substituting all occurrences of the real variable x with the interval variable $\mathbf{x}$ produces a rational interval function $F(\mathbf{x})$, called the *natural* interval extension of f.

THEOREM 3.2. *Given a real-valued, rational function f and its natural interval extension F such that $F(\mathbf{x})$ is well-defined for some $\mathbf{x} \in \mathbb{IR}$, we have*

(1) $\quad \mathbf{z} \subseteq \mathbf{z}' \subseteq \mathbf{x} \Rightarrow F(\mathbf{z}) \subseteq F(\mathbf{z}'),$ $\qquad$ (*inclusion isotonicity*)

(2) $\quad R(f; \mathbf{x}) \subseteq F(\mathbf{x}).$ $\qquad\qquad\qquad$ (*range enclosure*)

Proof. We demand that $F(\mathbf{x})$ be well-defined simply to avoid domain violations as discussed in Section 2.3.5. Part (1) then follows by repeatedly invoking Theorem 2.5, while evaluating F. To see that part (2) follows, we note that f and F are related by the equality $F([x, x]) = f(x)$. Now, assume that $R(f; \mathbf{x}) \not\subseteq F(\mathbf{x})$. Then there is a $\zeta \in \mathbf{x}$ such that $f(\zeta) \in R(f; \mathbf{x})$ but $f(\zeta) \notin F(\mathbf{x})$. This, however, implies that $f(\zeta) = F([\zeta, \zeta]) \notin F(\mathbf{x})$, which violates part (1). Hence $R(f; \mathbf{x}) \subseteq F(\mathbf{x})$. $\square$

We now want to treat more general functions than the rational ones. Monotone functions are easily extended: it suffices to evaluate the endpoints of the interval argument. To illustrate this, we may define

$$e^{\mathbf{x}} = \exp \mathbf{x} \ = [e^{\underline{x}}, e^{\overline{x}}]$$
$$\sqrt{\mathbf{x}} = \text{sqrt } \mathbf{x} \ = [\sqrt{\underline{x}}, \sqrt{\overline{x}}] \qquad \text{if } 0 \le \underline{x}$$
$$\log \mathbf{x} \ = [\log \underline{x}, \log \overline{x}] \qquad \text{if } 0 < \underline{x}$$
$$\arctan \mathbf{x} = [\arctan \underline{x}, \arctan \overline{x}].$$

Simple, non-monotone functions are also easily handled if they only have a finite number of extrema, which are known. As an example, we have

$$\mathbf{x}^n = \text{pow}(\mathbf{x}, n) = \begin{cases} [\underline{x}^n, \overline{x}^n] & : \text{if } n \in \mathbb{Z}^+ \text{ is odd,} \\ [\text{mig}(\mathbf{x})^n, \text{mag}(\mathbf{x})^n] & : \text{if } n \in \mathbb{Z}^+ \text{ is even,} \\ [1, 1] & : \text{if } n = 0, \\ [1/\overline{x}, 1/\underline{x}]^{-n} & : \text{if } n \in \mathbb{Z}^- \text{ and } 0 \notin \mathbf{x}. \end{cases}$$

Note that this definition provides tighter enclosures than the naive approach $\mathbf{x} \times \mathbf{x} \times \cdots \times \mathbf{x}$ (with n factors). This is because basic interval arithmetic does not distinguish between variables and their domains. To illustrate this important point, let us consider the square, $f(x) = x^2$. If $\mathbf{x} = [-2, 3]$, then $\mathbf{x} \times \mathbf{x} = [-6, 9]$, whereas $R(x^2; [-2, 3]) = [0, 9]$. Our definition of the power of an interval, however, recognizes that there is only one variable (and not n variables with coinciding domains) and thus produces a *sharp* enclosure of the range, that is, $R(x^n; \mathbf{a}) = \mathbf{a}^n$. Naturally, we always have $\mathbf{a}^n \subseteq \mathbf{a} \times \mathbf{a} \times \cdots \times \mathbf{a}$.

Trigonometric functions are handled in a similar manner. As an example, if we define $S^+ = \{2k\pi + \pi/2 : k \in \mathbb{Z}\}$ and $S^- = \{2k\pi - \pi/2 : k \in \mathbb{Z}\}$, we have

$$\sin \mathbf{x} =$$

$$\begin{cases} [-1, 1] & : \text{if } \mathbf{x} \cap S^- \neq [\emptyset] \text{ and } \mathbf{x} \cap S^+ \neq [\emptyset], \\ [-1, \max\{\sin \underline{x}, \sin \overline{x}\}] & : \text{if } \mathbf{x} \cap S^- \neq [\emptyset] \text{ and } \mathbf{x} \cap S^+ = [\emptyset], \\ [\min\{\sin \underline{x}, \sin \overline{x}\}, 1] & : \text{if } \mathbf{x} \cap S^- = [\emptyset] \text{ and } \mathbf{x} \cap S^+ \neq [\emptyset], \\ [\min\{\sin \underline{x}, \sin \overline{x}\}, \max\{\sin \underline{x}, \sin \overline{x}\}] & : \text{if } \mathbf{x} \cap S^- = [\emptyset] \text{ and } \mathbf{x} \cap S^+ = [\emptyset]. \end{cases}$$

Computing the interval version of $\sin x$ is thus reduced to determining whether the sets S^+ and S^- intersect the compact domain $\mathbf{x}$ or not, which is, in principle,[1] not hard. Note that if $\text{rad}(\mathbf{x}) \ge \pi$, then both sets have non-empty intersections with $\mathbf{x}$. Using standard identities such as $\cos x = \sin(x + \frac{\pi}{2})$, we can obtain interval extensions for all trigonometric functions in a similar fashion.

Exercise 3.3. *Compute* $\sin[\frac{\pi}{4}, \frac{3\pi}{4}]$, $\cos[\frac{-\pi}{4}, \frac{3\pi}{2}]$, *and* $\tan[-\frac{\pi}{4}, \frac{\pi}{4}]$.

Exercise 3.4. *Find the interval expression for* $\tan \mathbf{x}$, *just as we did for* $\sin \mathbf{x}$.

Exercise 3.5. *How would you define* $\mathbf{x}^\alpha$ *for non-integer exponents? Does this definition also work for interval exponents* α?

---

[1] An important implementation issue is that if the interval $\mathbf{x}$ is located very far away from the origin, then the knowledge of a large number of leading digits of π is required.

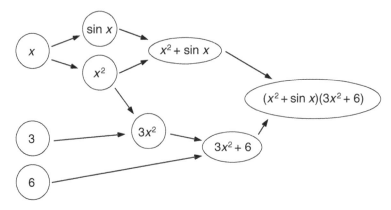

Figure 3.1 A directed acyclic graph for $f(x) = (x^2 + \sin x)(3\,x^2 + 6)$.

For future reference, we define the class of standard functions to be the set

$$\mathfrak{S} = \{a^x, \log_a x, x^{p/q}, \text{abs}\, x, \sin x, \cos x, \tan x, \dots$$

$$\dots, \sinh x, \cosh x, \tanh x, \arcsin x, \arccos x, \arctan x\}.$$

By methods similar to those just described, it is possible to extend all standard functions to the interval realm: any $f \in \mathfrak{S}$ has a sharp interval extension F. Again, by sharp, we mean that the interval evaluation $F(\boldsymbol{x})$ produces the exact range of f over the domain $\boldsymbol{x}$:

$$f \in \mathfrak{S} \Rightarrow R(f; \boldsymbol{x}) = F(\boldsymbol{x}).$$

Note that, in particular, this implies that $F([x, x]) = f(x)$, that is, F and f are identical on $\mathbb{R}$.

Of course, the class of standard functions is too small for most practical applications. We will use them as building blocks for more complicated functions as follows.

Definition 3.6. *Any real-valued function expressed as a finite number of standard functions combined with constants, arithmetic operations, and compositions is called an elementary function. The class of elementary functions is denoted by* $\mathfrak{E}$.

So by definition, an elementary function can be expressed as a finite number of operations. This (non-unique) way of representing the function in terms of its most basic sub-expressions produces an explicit means of coding the function. The resulting code can be represented as Directed Acyclic Graph (DAG); see Figure 3.1.

Unfortunately, we cannot obtain sharp extensions of all elementary functions. This is made clear by even very simple examples such as the following:

Example 3.1.1 *Let* $f : \mathbb{R} \to \mathbb{R}$ *be defined by* $f(x) = \frac{x}{1+x^2}$. *Its natural interval extension is given by* $F(\boldsymbol{x}) = \frac{\boldsymbol{x}}{1+\boldsymbol{x}^2}$. *Thus, for* $\boldsymbol{x} = [1, 2]$, *we have*

$$F(\boldsymbol{x}) = F([1, 2]) = \frac{[1, 2]}{1 + [1, 2]^2} = \frac{[1, 2]}{[2, 5]} = [\tfrac{1}{5}, 1],$$

whereas $R(f; [1, 2]) = [\tfrac{2}{5}, \tfrac{1}{2}]$ *(prove this). Hence,* $R(f; [1, 2]) \overset{\circ}{\subset} F([1, 2])$.

In general, we cannot hope for a sharp extension if the variable x occurs more than once in the explicit representation of f. It should also be pointed out that an elementary function f has infinitely many interval extensions: if F is an extension of f, then so is $F(\mathbf{x}) + \mathbf{x} - \mathbf{x}$. Given an explicit representation of an elementary function f, we say that the extension obtained by replacing all occurrences of x by the interval $\mathbf{x}$ is the *natural* extension. Although the representation of f is immaterial when computing over $\mathbb{R}$, it *does*[2] make a big difference over $\mathbb{IR}$.

Example 3.1.2 *Let* $f_1(x) = 1 - x^2$, $f_2(x) = 1 - x \cdot x$, *and* $f_3(x) = (1 - x)(1 + x)$. *These three functions are indistinguishable when evaluated over* $\mathbb{R}$. *The corresponding natural interval extensions, however, will be distinct functions over* $\mathbb{IR}$. *They are* $F_1(\mathbf{x}) = 1 - \mathbf{x}^2$, $F_2(\mathbf{x}) = 1 - \mathbf{x} \times \mathbf{x}$, *and* $F_3(\mathbf{x}) = (1 - \mathbf{x})(1 + \mathbf{x})$. *Note that* $F_1 \neq F_2 \neq F_3$ *over* $\mathbb{IR}$. *This is easily seen by evaluating the interval functions at, for example, the interval* $[-1, 1]$.

$$F_1([-1, 1]) = 1 - [-1, 1]^2 = [1, 1] - [0, 1] = [0, 1],$$

$$F_2([-1, 1]) = 1 - [-1, 1] \times [-1, 1] = [1, 1] - [-1, 1] = [0, 2],$$

$$F_3([-1, 1]) = (1 - [-1, 1])(1 + [-1, 1]) = [0, 2] \times [0, 2] = [0, 4].$$

Therefore, we distinguish between different representations of the same elementary function; each explicit representation is considered as a distinct member of $\mathfrak{E}$.

Somewhat more subtle is the fact that some perfectly valid elementary functions do not admit well-defined natural interval extensions.

Example 3.1.3 *Let* $f : \mathbb{R} \to \mathbb{R}$ *be defined by* $f(x) = \frac{\sin \pi x}{x}$ *for* $x \neq 0$ *and* $f(0) = \pi$. *Then* f *is a continuous function on* $\mathbb{R}$. *It is a simple exercise to show that* $R(f; [-1, 1]) = [0, \pi]$, *yet the natural interval extension of* f *will be undefined for any interval containing the origin due to the (removable) singularity.*

Later on, we will learn several ways of handling such functions gracefully.[3]

Although the elementary functions do not, in general, admit sharp interval extensions, it is possible to prove that the extensions, when well-defined, are *inclusion isotonic*.

DEFINITION 3.7. *Let* $\mathbf{x} \in \mathbb{IR}$. *An interval-valued function* $F : \mathbf{x} \cap \mathbb{IR} \to \mathbb{IR}$ *is inclusion isotonic if, for all* $\mathbf{z} \subseteq \mathbf{z}' \subseteq \mathbf{x}$, *we have* $F(\mathbf{z}) \subseteq F(\mathbf{z}')$.

We make this statement precise in part (1) of the following theorem:

THEOREM 3.8. (The Fundamental Theorem of Interval Analysis) *Given an elementary function* f *and a natural interval extension* F *such that* $F(\mathbf{x})$ *is well-defined for some* $\mathbf{x} \in \mathbb{IR}$, *we have*

$$(1) \qquad \mathbf{z} \subseteq \mathbf{z}' \subseteq \mathbf{x} \Rightarrow F(\mathbf{z}) \subseteq F(\mathbf{z}'), \qquad \text{(inclusion isotonicity)}$$

$$(2) \qquad R(f; \mathbf{x}) \subseteq F(\mathbf{x}). \qquad \text{(range enclosure)}$$

---

[2] In this sense, computing over $\mathbb{IR}$ and $\mathbb{F}$ is similar: in both situations, the explicit representation of a function is relevant.

[3] Again, computing over $\mathbb{IR}$ and $\mathbb{F}$ is similar here: in both situations, any naive attempt to evaluate a function at a removable singularity will fail.

Proof. Recall that an elementary function is recursively defined in terms of its sub-expressions. The theorem clearly holds for rational functions (see Theorem 3.2), as well as for the standard functions (since they are sharp). Thus it suffices to prove that if the theorem holds for the elementary functions g_1 and g_2, then it also holds for $g_1 \star g_2$, where $\star \in \{+, -, \times, \div, \circ\}$. We will prove the claim for the composition operator $\circ$; the remaining cases are completely analogous. Since $F(x)$ is well-defined, neither the real-valued function f nor its sub-expressions g_i have singularities in their domains induced by the set x. In particular, g_2 is continuous on any subintervals $w \subseteq w'$ of its domain. This means that $G_2(w)$ and $G_2(w')$ are compact intervals: y and y', respectively. Therefore, since G_1 and G_2 are inclusion isotonic, we have $y \subseteq y'$ and

$$G_1 \circ G_2(w) = G_1(y) \subseteq G_1(y') = G_1 \circ G_2(w').$$

Part (2) now follows immediately by the same argument as given in the proof of Theorem 3.2.   □

The crucial implication of Theorem 3.8 is the fact that we have a means of bounding the range of a function by considering its interval extension. Although the exact range $R(f; x)$ is virtually impossible to compute, its enclosure $F(x)$ is easy to come by.

Forming the contra-positive of the statement "$y \in R(f; x) \Rightarrow y \in F(x)$" produces the very useful "$y \notin F(x) \Rightarrow y \notin R(f; x)$." This can be used, for example, when searching for roots of a function f: if $0 \notin F(x)$, then we know that f has no roots in the domain x. This can be used for discarding portions of a larger initial search space.

Example 3.1.4 *Let $f(x) = (\sin x - x^2 + 1) \cos x$. Prove that f has no roots in the compact interval $x = [0, \frac{1}{2}]$.*

The most common way to approach this problem is to find all critical points of f within x. These are the points for which $f'(x) = 0$. Once they are obtained, we evaluate f at these points, as well as at the endpoints of x. If all computed function values have the same sign, we can safely conclude that f has no roots in x. Note, however, that we are simply substituting one root-finding problem ($f(x) = 0$) for another ($f'(x) = 0$). It may very well be the case that f' is more complicated than f, in which case we will have gained nothing.

Using interval techniques, however, we can obtain a proof in just one evaluation of the interval extension of f:

$$F([0, \tfrac{1}{2}]) = (\sin [0, \tfrac{1}{2}] - [0, \tfrac{1}{2}]^2 + 1) \cos [0, \tfrac{1}{2}]$$

$$= ([0, \sin \tfrac{1}{2}] - [0, \tfrac{1}{4}] + 1) \times [\cos \tfrac{1}{2}, 1]$$

$$= ([0, \sin \tfrac{1}{2}] + [\tfrac{3}{4}, 1]) \times [\cos \tfrac{1}{2}, 1]$$

$$= [\tfrac{3}{4}, 1 + \sin \tfrac{1}{2}] \times [\cos \tfrac{1}{2}, 1]$$

$$= [\tfrac{3}{4} \cos \tfrac{1}{2}, 1 + \sin \tfrac{1}{2}] \subseteq [0.65818, 1.4795].$$

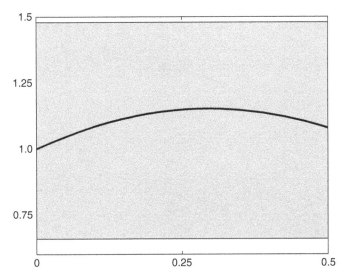

Figure 3.2 The function $f(x) = (\sin x - x^2 + 1)\cos x$ and its interval enclosure over the
domain $[0, \frac{1}{2}]$.

*Now, since $0 \notin F([0, \frac{1}{2}])$, it follows that $0 \notin R(f; [0, \frac{1}{2}])$, so f has no roots in the
interval $[0, \frac{1}{2}]$ (see Figure 3.2).*

Here we were fortunate: with just one interval evaluation, the range could be
enclosed sufficiently tight for our needs. Occasionally, we will not get *any* information with just one interval evaluation. This happens when $F(\mathbf{x})$ is not well-defined,
which can occur even though $R(f; \mathbf{x})$ is a compact interval. The following example illustrates that when this happens, we may invoke part (1) of Theorem 3.8 to
obtain a more satisfactory result.

Example 3.1.5 *Find an enclosure of $R(f; [0, 2])$ where $f(x) = \sqrt{x + \sin 2x}$.
This time, a single evaluation of the interval extension of f will not suffice:*

$$R(f; [0, 2]) \subseteq F([0, 2]) = \sqrt{[0, 2] + \sin(2 \times [0, 2])} = \sqrt{[0, 2] + \sin[0, 4]}$$

$$= \sqrt{[0, 2] + [\sin 4, \sin \pi/2]} = \sqrt{[0, 2] + [\sin 4, 1]}$$

$$= \sqrt{[\sin 4, 3]}.$$

*Note that the final expression cannot be properly evaluated since $\sin 4$ is negative.[4]
If we bisect the original domain $[0, 2] = [0, 1] \cup [1, 2]$, we can use the inclusion*

---

[4]One can, as in Section 2.3.5, also propose the interpretation $\sqrt{\mathbf{x}} = \sqrt{\mathbf{x} \cap [0, +\infty]}$, which resolves
this issue.

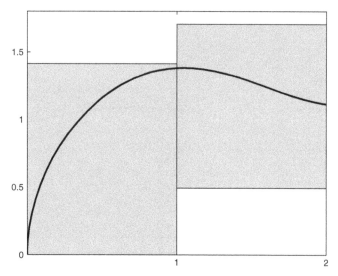

Figure 3.3 The function $f(x) = \sqrt{x + \sin 2x}$ and its interval enclosure over the domain $[0, 2]$.

isotonicity of F to obtain

$$R(f; [0, 2]) = R(f; [0, 1]) \cup R(f; [1, 2]) \subseteq F([0, 1]) \cup F([1, 2])$$

$$= \sqrt{[0, 1] + \sin(2 \times [0, 1])} \cup \sqrt{[1, 2] + \sin(2 \times [1, 2])}$$

$$= \sqrt{[0, 1] + \sin[0, 2]} \cup \sqrt{[1, 2] + \sin[2, 4]}$$

$$= \sqrt{[0, 1] + [\sin 0, \sin \tfrac{\pi}{2}]} \cup \sqrt{[1, 2] + [\sin 4, \sin 2]}$$

$$= \sqrt{[0, 1] + [0, 1]} \cup \sqrt{[1 + \sin 4, 2 + \sin 2]}$$

$$= [0, \sqrt{2}] \cup [\sqrt{1 + \sin 4}, \sqrt{2 + \sin 2}]$$

$$= [0, \sqrt{2 + \sin 2}] \subseteq [0, 1.7057].$$

This produces a valid (non-sharp) enclosure of $R(f; [0, 2])$ (see Figure 3.3).

Of course, the enclosure $F(x)$ may grossly overestimate $R(f; x)$. If f is sufficiently regular, however, this overestimation can be made arbitrarily small by subdividing x into many smaller pieces, evaluating F over each smaller piece, and then taking the union of all resulting sets.

DEFINITION 3.9. *A function $f: D \to \mathbb{R}$ is Lipschitz if there exists a positive constant K such that, for all $x, y \in D$, we have $|f(x) - f(y)| \le K|x - y|$. The number K is called a Lipschitz constant of f over D.*

Thus a Lipschitz function f is necessarily continuous, but it need not be differentiable. If f happens to be differentiable, then the modulus of the derivative is always bounded by the Lipschitz constant K.

We now define $\mathfrak{E}_{\mathfrak{L}}$ to be the set of all elementary functions whose sub-expressions are all Lipschitz:

$$\mathfrak{E}_{\mathfrak{L}} = \{f \in \mathfrak{E} : \text{each sub-expression of } f \text{ is Lipschitz}\}.$$

THEOREM 3.10. (Range Enclosure) *Consider* $f : I \rightarrow \mathbb{R}$ *with* $f \in \mathfrak{E}_{\mathfrak{L}}$, *and let* F *be an inclusion isotonic interval extension of* f *such that* $F(\mathbf{x})$ *is well-defined for some* $\mathbf{x} \subseteq I$. *Then there exists a positive real number* K, *depending on* F *and* $\mathbf{x}$, *such that, if* $\mathbf{x} = \cup_{i=1}^{k} \mathbf{x}^{(i)}$, *then*

$$R(f; \mathbf{x}) \subseteq \bigcup_{i=1}^{k} F(\mathbf{x}^{(i)}) \subseteq F(\mathbf{x})$$

and

$$\text{rad}\left(\bigcup_{i=1}^{k} F(\mathbf{x}^{(i)})\right) \leq \text{rad}(R(f; \mathbf{x})) + K \max_{i=1,\dots,k} \text{rad}\left(\mathbf{x}^{(i)}\right).$$

Proof. We start with the first inclusion. By the the inclusion isotonic property, it follows that by splitting the box $\mathbf{x}$ into smaller pieces $\mathbf{x} = \cup_{i=1}^{k} \mathbf{x}^{(i)}$, we have

$$R(f; \mathbf{x}) = R(f; \cup_{i=1}^{k} \mathbf{x}^{(i)}) = \bigcup_{i=1}^{k} R(f; \mathbf{x}^{(i)}) \subseteq \bigcup_{i=1}^{k} F(\mathbf{x}^{(i)}) \subseteq F(\cup_{i=1}^{k} \mathbf{x}^{(i)}) = F(\mathbf{x}).$$

To prove the second statement, we must show that if $\tilde{\xi} \in \cup_{i=1}^{k} F(\mathbf{x}^{(i)})$, then there exists $\xi \in R(f; \mathbf{x})$ such that $|\tilde{\xi} - \xi| \leq K \max_{i=1,\dots,k} \text{rad}\left(\mathbf{x}^{(i)}\right)$. We will prove the more precise statement that if $\mathbf{w} \subseteq \mathbf{x}$ and $\tilde{\xi} \in F(\mathbf{w})$, then for all $\xi \in R(f; \mathbf{w})$ we have $|\tilde{\xi} - \xi| \leq K \text{rad}(\mathbf{w})$. This clearly implies the second part of the theorem.

We first note that the statement is trivially valid for constants and standard functions, since they return sharp enclosures (which are thus bounded). Therefore it suffices to show that if the statement is true for two connecting branches g_1 and g_2 of the recursive tree for f, then it also holds for $g_1 \star g_2$, where $\star \in \{+, -, \times, \div, \circ\}$. In a sub-expression like $\sin x^2$, we interpret $\sin$ and x^2 as connecting branches. As in the proof of Theorem 3.8, we will only prove the claim for the composition operator $\circ$, as the remaining cases are very similar.

Since f is an elementary function, g_1 and g_2 are also elementary, so their interval extensions G_1 and G_2 are inclusion isotonic by Theorem 3.8. Furthermore, since $f \in \mathfrak{E}_{\mathfrak{L}}$, g_1 and g_2 are Lipschitz on their domains, and since $F(\mathbf{x})$ is well-defined, the extensions G_1 and G_2 are also well-defined on their domains ($\mathbf{w}_1$ resp. $\mathbf{w}_2$) induced by $\mathbf{x}$.

Our inductive assumptions are

$$\mathbf{v} \subseteq \mathbf{w}_i, \tilde{z} \in G_i(\mathbf{v}), \text{ and } z \in R(g_i; \mathbf{v}) \Rightarrow |\tilde{z} - z| \leq K_i \text{rad}(\mathbf{v}). \qquad (i = 1, 2)$$

From part (2) of Theorem 3.8, we have (for all $\mathbf{v} \subseteq \mathbf{w}_2$)

$$R(g_1 \circ g_2; \mathbf{v}) = R(g_1; R(g_2; \mathbf{v})) \subseteq R(g_1; G_2(\mathbf{v})).$$

Now, if $z \in R(g_1 \circ g_2; \mathbf{v})$, then there exists $u \in R(g_2; \mathbf{v})$ such that $z = g_1(u)$. Note that $u \in G_2(\mathbf{v})$. Therefore, if $\tilde{z} \in G_1 \circ G_2(\mathbf{v}) = G_1(G_2(\mathbf{v}))$, then by our assumptions on g_1 and G_1, it follows that

$$|z - \tilde{z}| \le K_1 \text{rad}(G_2(\mathbf{v})).$$

Now, by our assumptions on g_2 and G_2, it follows that

$$\text{rad}(G_2(\mathbf{v})) \le \text{rad}(R(g_2; \mathbf{v})) + K_2 \text{rad}(\mathbf{v}) \le K_3 \text{rad}(\mathbf{v}) + K_2 \text{rad}(\mathbf{v}),$$

where we used the Lipschitz property of g_2 in the final estimate. Combining these two inequalities yields

$$|z - \tilde{z}| \le K_1(K_3 + K_2)\text{rad}(\mathbf{v}).$$

Since f is elementary, its recursive tree is finite, and thus the accumulated constants from successive estimates as above will be finite. This finishes the proof. □

In essence, the second part of Theorem 3.10 says that if the listed conditions are satisfied, then the overestimation tends to zero no slower than linearly as the domain shrinks:

$$\text{rad}(\mathbf{x}) = \mathcal{O}(\varepsilon) \Rightarrow d(R(f; \mathbf{x}), F(\mathbf{x})) = \mathcal{O}(\varepsilon),$$

where $d(\cdot, \cdot)$ is the Hausdorff distance, as defined in (2.1). Since Lipschitz functions satisfy $\text{rad}(R(f; \mathbf{x})) = \mathcal{O}(\text{rad}(\mathbf{x}))$, it also follows that

$$\text{rad}(\mathbf{x}) = \mathcal{O}(\varepsilon) \Rightarrow \text{rad}(F(\mathbf{x})) = \mathcal{O}(\varepsilon),$$

that is, the width of the enclosure scales at least linearly with ε.

Exercise 3.11. *Prove that if f is Lipschitz with Lipschitz constant K, then*

$$\text{rad}(R(f; \mathbf{x})) \le K \text{rad}(\mathbf{x}).$$

Example 3.1.6 *Let $f(x) = (\sin x - x^2 + 1) \cos x$ be as in Example 3.1.5, and consider the domain $[0, \varepsilon]$, where ε is small. By a direct computation we have*

$$R(f; [0, \varepsilon]) = [1, (\sin \varepsilon - \varepsilon^2 + 1) \cos \varepsilon],$$
$$F([0, \varepsilon]) = [(1 - \varepsilon^2) \cos \varepsilon, 1 + \sin \varepsilon].$$

The overestimation is therefore given by the distance between the two intervals:

$$d(R(f; [0, \varepsilon]), F([0, \varepsilon]))$$
$$= \max \left\{ |1 - (1 - \varepsilon^2) \cos \varepsilon|, |(\sin \varepsilon - \varepsilon^2 + 1) \cos \varepsilon - (1 + \sin \varepsilon)| \right\}$$
$$= \tfrac{3}{2}\varepsilon^2 + \mathcal{O}(\varepsilon^3) = \mathcal{O}(\varepsilon^2)$$

Note that this is actually better than claimed by the theorem: the overestimation is quadratic in ε rather than being linear.

If we consider functions that are not Lipschitz, then the results of the theorem may not hold.

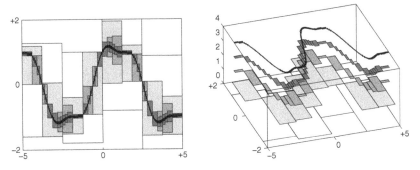

Figure 3.4 Two views of the function $f(x) = \cos^3 x + \sin x$ and its successively tighter interval enclosures over the domain $[-5, 5]$.

Example 3.1.7 *Let $f(x) = \sqrt{x} + \sin 2x$ be as in Example 3.1.5, and consider the domain $[0, \varepsilon]$, where ε is small. Again, by a direct computation we have*

$$R(f; [0, \varepsilon]) = [0, \sqrt{\varepsilon + \sin 2\varepsilon}] = F([0, \varepsilon]).$$

In this particular case, there is no overestimation at all. Nevertheless, if we measure the width of the enclosure, we have

$$\mathrm{rad}\big(F([0, \varepsilon])\big) = \frac{1}{2}\sqrt{\varepsilon + \sin 2\varepsilon} = \mathcal{O}(\sqrt{\varepsilon}) \neq \mathcal{O}(\varepsilon).$$

This "slower than linear" scaling is due to the fact that f' has a singularity at the origin.

As a consequence of Theorem 3.10, we can generate tight enclosures of the graph of a given function. By partitioning the domain into sufficiently small subintervals, we can obtain the desired accuracy, that is, the overestimation can be made as small as we wish. In Figure 3.4, this is illustrated for the function $f(x) = \cos^3 x + \sin x$ over the domain $[-5, 5]$. By setting the tolerance in the range width (i.e., the height of the boxes) to 2, 1, 0.5, and 0.25, we get 4, 8, 22, and 40 enclosing boxes, respectively.

3.2 CENTERED FORMS

In some situations, we can arrange for the overestimation of the exact range to tend to zero faster than linearly as the domain shrinks. This is possible if the function f satisfies the Mean Value Theorem.

THEOREM 3.12. (The Mean Value Theorem) *If f is continuous on $[a, b]$, and differentiable on (a, b), then there is a number ζ in (a, b) such that*

$$f'(\zeta) = \frac{f(b) - f(a)}{b - a}.$$

For the remainder of this section, we will only consider functions satisfying the assumptions of the Mean Value Theorem.

Thus, given $f : \mathbf{x} \to \mathbb{R}$, the Mean Value Theorem states that if x and c are points in $\mathbf{x}$, then there is a ζ between x and c (and thus also in $\mathbf{x}$) such that

$$f(x) = f(c) + f'(\zeta)(x - c).$$

Now suppose that we have an interval extension F' for the derivative f'. It then follows that

$$f(x) = f(c) + f'(\zeta)(x - c) \in f(c) + F'(\mathbf{x})(x - c) \subseteq f(c) + F'(\mathbf{x})(\mathbf{x} - c),$$

where the last expression is independent of both x and ζ. Thus, for all x and c in the domain $\mathbf{x}$, we have

$$R(f; \mathbf{x}) \subseteq f(c) + F'(\mathbf{x})(\mathbf{x} - c) \stackrel{\text{def}}{=} F(\mathbf{x}; c).$$

The interval function $F(\mathbf{x}; c)$ is called a *centered form*.[5] The most popular choice is to take $c = \mathrm{mid}(\mathbf{x})$, which produces the *mean-value form*,

$$F_m(\mathbf{x}) \stackrel{\text{def}}{=} F(\mathbf{x}; \mathrm{mid}(\mathbf{x})).$$

Using the shorthand notations $m = \mathrm{mid}(\mathbf{x})$ and $r = \mathrm{rad}(\mathbf{x})$, we have

$$F_m(\mathbf{x}) = f(m) + F'(\mathbf{x})(\mathbf{x} - m) = f(m) + F'(\mathbf{x})[-r, r].$$

One of the benefits of using centered forms is that they provide us with *explicit* bounds on the overestimation of the exact range, as opposed to Theorem 3.10.

THEOREM 3.13. (Centered Form Enclosure) *Consider an elementary function* $f : I \to \mathbb{R}$ *satisfying the Mean Value Theorem, and let F' be an interval extension of f' such that $F'(\mathbf{x})$ is well-defined for some $\mathbf{x} \subseteq I$. Then, if c belongs to $\mathbf{x}$, we have*

$$R(f; \mathbf{x}) \subseteq F(\mathbf{x}; c)$$

and

$$\mathrm{rad}\big(F(\mathbf{x}; c)\big) \leq \mathrm{rad}\big(R(f; \mathbf{x})\big) + 4\mathrm{rad}\big(F'(\mathbf{x})\big)\mathrm{rad}\big(\mathbf{x}\big).$$

Proof. We have already proved the inclusion $R(f; \mathbf{x}) \subseteq F(\mathbf{x}; c)$, so we need only concentrate our efforts toward proving the bound on the overestimation.

Let y be any element of $F(\mathbf{x}; c)$. Then there is a $u \in F'(\mathbf{x})$ and a $v \in \mathbf{x} - c$ such that $y = f(c) + uv$. Now, $v = x - c$ for some $x \in \mathbf{x}$, so

$$y = f(c) + u(x - c). \tag{3.1}$$

By the Mean Value Theorem, we have $f(x) = f(c) + f'(\zeta)(x - c)$ for some $\zeta \in \mathbf{x}$, that is,

$$f(c) = f(x) - f'(\zeta)(x - c). \tag{3.2}$$

Combining (3.1) and (3.2) produces

$$y = f(x) - f'(\zeta)(x - c) + u(x - c)$$

$$= f(x) + (u - f'(\zeta))(x - c) \in R(f; \mathbf{x}) + \big(F'(\mathbf{x}) - F'(\mathbf{x})\big)(\mathbf{x} - c).$$

---

[5]When working in finite precision, it is crucial to compute $F([c, c])$ rather than $f(c)$ in order to take all rounding errors into account.

Since y was an arbitrary element of $F(x; c)$, it follows that the overestimation is bounded by the size of the set

$$\big(F'(x) - F'(x)\big)(x - c). \tag{3.3}$$

Note that the first factor of (3.3) is an interval of the form $a = \lambda \cdot [-1, 1]$. If b is an arbitrary interval, we have a particularly elegant formula for the product:

$$a \times b = \lambda \cdot \text{mag}(b) \cdot [-1, 1].$$

Setting $a = F'(x) - F'(x)$ and $b = x - c$ produces $\lambda = 2\text{rad}\big(F'(x)\big)$ and $\text{mag}(b) \leq 2\text{rad}(x)$, that is,

$$\big(F'(x) - F'(x)\big)(x - c) \subseteq 4\text{rad}\big(F'(x)\big)\text{rad}(x) \cdot [-1, 1].$$

Thus we have established the desired bound

$$\text{rad}\big(\big(F'(x) - F'(x)\big)(x - c)\big) \leq 4\text{rad}\big(F'(x)\big)\text{rad}(x),$$

which concludes the proof. □

We remark that the bound can be improved by a factor of two if the mean-value form is used rather than a general centered form. This is because we then have $b = x - \text{mid}(x) = \text{rad}(x) \cdot [-1, 1]$, which gives the improved bound $\text{mag}(b) = \text{rad}(x)$.

Exercise 3.14. *Prove that a centered form $F(x, c)$ is not inclusion isotonic unless $c = \text{mid}(x)$.*

It is clear from Theorem 3.13 that we obtain a very tight enclosure of $R(f; x)$ when both $F'(x)$ and x are narrow intervals. If f' satisfies the conditions of Theorem 3.10, then $\text{rad}(x) = \mathcal{O}(\varepsilon) \Rightarrow \text{rad}\big(F'(x)\big) = \mathcal{O}(\varepsilon)$, and the overestimation is roughly of the order $\mathcal{O}(\varepsilon^2)$. Nevertheless, it does not follow that centered forms always produce better enclosures than a single interval evaluation.

Example 3.2.1 *Let $f(x) = \frac{x^2+1}{x}$. Then $f'(x) = \frac{x^2-1}{x^2}$, so $F'(x)$ is well-defined as long as $0 \notin x$. Let $x = [1, 2]$. Then the interval evaluation gives*

$$F([1, 2]) = \frac{[1, 2]^2 + 1}{[1, 2]} = \frac{[2, 5]}{[1, 2]} = [1, 5],$$

whereas the mean-value form produces

$$F_m([1, 2]) = f(\tfrac{3}{2}) + F'([1, 2])[-\tfrac{1}{2}, \tfrac{1}{2}]$$

$$= \tfrac{13}{6} + \frac{[1, 2]^2 - 1}{[1, 2]^2}[-\tfrac{1}{2}, \tfrac{1}{2}] = \tfrac{13}{6} + \frac{[0, 3]}{[1, 4]}[-\tfrac{1}{2}, \tfrac{1}{2}]$$

$$= \tfrac{13}{6} + [0, 3][-\tfrac{1}{2}, \tfrac{1}{2}] = \tfrac{13}{6} + [-\tfrac{3}{2}, \tfrac{3}{2}] = [\tfrac{2}{3}, \tfrac{11}{3}].$$

Thus $\text{rad}(F_m(x)) = \tfrac{3}{2} < 2 = \text{rad}(F(x))$, so the mean-value form gives the more narrow enclosure. In spite of this, the interval evaluation produces a better lower bound for the exact range. Intersecting both enclosures produces the even tighter enclosure:

$$R(f; [1, 2]) \subseteq F([1, 2]) \cap F_m([1, 2]) = [1, \tfrac{11}{3}].$$

Example 3.2.2 *Let* $f(x) = \sin x^2$. *Then* $f'(x) = 2x \cos x^2$, *so if we introduce* $\mathbf{x} = m + [-r, r]$, *then we have*

$$F_m(\mathbf{x}) = f(m) + F'(\mathbf{x})[-r, r] = \sin m^2 + 2\mathbf{x} \cos (\mathbf{x}^2)[-r, r].$$

For $\mathbf{x} = [-\varepsilon, \varepsilon]$, *we have* $m = 0$ *and* $r = \varepsilon$, *which gives*

$$F_m([-\varepsilon, \varepsilon]) = \sin 0^2 + 2[-\varepsilon, \varepsilon] \cos \left([0, \varepsilon^2]\right)[-\varepsilon, \varepsilon]$$
$$= 2[-\varepsilon^2, \varepsilon^2][\cos \varepsilon^2, 1] = 2[-\varepsilon^2, \varepsilon^2].$$

The interval evaluation, however, produces the sharp enclosure:

$$F([-\varepsilon, \varepsilon]) = \sin [-\varepsilon, \varepsilon]^2 = \sin [0, \varepsilon^2] = [0, \sin \varepsilon^2] = R(f; [-\varepsilon, \varepsilon]).$$

In this particular case, the mean-value form produces an enclosure that is more than four times wider than the exact range:

$$\mathrm{rad}\big(F(\mathbf{x})\big) \leq \tfrac{1}{4}\mathrm{rad}\big(F_m(\mathbf{x})\big).$$

3.3 MONOTONICITY

When computing a centered form, the quantity $F'(\mathbf{x})$ must be calculated. If it happens that $0 \notin F'(\mathbf{x})$, then f is monotone on $\mathbf{x}$, and a sharp result can easily be obtained:

$$R(f; \mathbf{x}) = \begin{cases} [f(\underline{x}), f(\overline{x})] & \text{if } \min\{y \in F'(\mathbf{x})\} \geq 0, \\ [f(\overline{x}), f(\underline{x})] & \text{if } \max\{y \in F'(\mathbf{x})\} \leq 0. \end{cases}$$

Example 3.3.1 *Let us return to Example 3.2.1, where* $f(x) = \frac{x^2+1}{x}$ *is considered on the domain* $\mathbf{x} = [1, 2]$. *Then* $f'(x) = \frac{x^2-1}{x^2}$, *so*

$$F'([1, 2]) = \frac{[1, 2]^2 - 1}{[1, 2]^2} = \frac{[1, 4] - 1}{[1, 4]} = \frac{[0, 3]}{[1, 4]} = [0, 3].$$

This means that f *is non-decreasing on* $[1, 2]$, *so the exact range is given by evaluating the endpoints:*

$$R(f; [1, 2]) = [f(1), f(2)] = [2, \tfrac{5}{2}].$$

If we compare the widths resulting from the various approaches, we have

$$\mathrm{rad}\big(R(f; [1, 2])\big) = \frac{1}{4} < \mathrm{rad}\big(F_m(\mathbf{x})\big) = \frac{3}{2} < \mathrm{rad}\big(F(\mathbf{x})\big) = 2.$$

In Chapter 4, we will present a very powerful technique that allows us to generate graph enclosures of any desired order. This can be used to obtain very tight enclosures of the range.

3.4 COMPUTER LAB II

Problem 1. Write a small set of routines supporting interval versions of the standard functions e^x, $\log x$, x^n ($n \in \mathbb{Z}$), x^α ($\alpha \in \mathbb{R} \setminus \mathbb{Z}$), $\sin x$, $\cos x$, and $\tan x$. If you have access to directed rounding, use it. You may assume that the floating point implementation of each standard function returns the floating point nearest the exact value.

Problem 2. Merge the routines from the previous problem with the routines from Problem 6 in Computer Lab I. Use this to compute $f(1 + 2^{-k}[-1, 1])$ for $k = 0, \dots, 10$, where

$$f(x) = e^{\sin e^{\cos x + 2x^5}},$$

or in a more functional form:

$$f(x) = \exp(\sin(\exp(\cos(x) + 2 * \mathrm{pow}(x, 5)))).$$

Problem 3. Let f be as in Problem 2 above. Compute (by hand?) the formal expression for $f'(x)$ and implement the mean-value form $F_m(x)$. Use this to compute $f(1 + 2^{-k}[-1, 1])$ for $k = 0, \dots, 10$. Compare the quality of the enclosures with those from Problem 2.

Chapter Four

Automatic Differentiation

IN THIS CHAPTER, we will present an elegant and effective technique for automatically generating n-th order derivatives of a given function. This will open the path to a series of powerful interval methods, almost completely removing the main obstruction of overestimation seen in the previous chapters. As a nice by-product, the technique also relieves us from the tedious and error-prone task of hand-coding functional expressions for derivatives. A thorough treatment of the topic at hand can be found in [Gr00].

4.1 FIRST-ORDER DERIVATIVES

The aim of this section is to produce the value $f'(x_0)$, given the point x_0 and a formula for f. This is usually achieved by either computing an approximation of $f'(x_0)$ via one of the divided differences

$$\Delta_h^+(f, x_0) \overset{\text{def}}{=} \frac{f(x_0 + h) - f(x_0)}{h},$$

$$\Delta_h^-(f, x_0) \overset{\text{def}}{=} \frac{f(x_0) - f(x_0 - h)}{h}, \tag{4.1}$$

$$\Delta_h^\pm(f, x_0) \overset{\text{def}}{=} \frac{f(x_0 + h) - f(x_0 - h)}{2h},$$

or by generating the exact symbolic formula for f' and then evaluating it at x_0. Both approaches have serious disadvantages. The former is prone to producing gross errors due to both the discretization and the finite precision computations. The latter approach is both memory- and time-consuming, and many functions f simply cannot be handled in a reasonably short period of time.

We will, much like with the interval arithmetic, start from scratch and construct a *differentiation arithmetic*. Focusing initially on the real-valued setting, we will perform all calculations with ordered pairs of real numbers

$$\vec{u} = (u, u'),$$

where u denotes the value of the function $u : \mathbb{R} \to \mathbb{R}$ evaluated at the point x_0, and where u' denotes the value $u'(x_0)$. The basic arithmetic rules are

$$\vec{u} + \vec{v} = (u + v, u' + v')$$

$$\vec{u} - \vec{v} = (u - v, u' - v') \tag{4.2}$$

$$\vec{u} \times \vec{v} = (uv, uv' + u'v)$$
$$\vec{u} \div \vec{v} = (u/v, (u' - (u/v)v')/v),$$

where we demand that $v \neq 0$ when dividing. The rule for division is derived from

$$\left(\frac{u}{v}\right)' = \frac{u'v - uv'}{v^2} = \frac{u' - (u/v)v'}{v}.$$

Note that the quantity (u/v) need only be computed once, although it appears twice in the formula for $\vec{u} \div \vec{v}$.

In order to be able to compute with this new arithmetic, we need to know how constants and the independent variable x are treated. Following the usual rules of differentiation, we define

$$\vec{x} = (x, 1) \quad \text{and} \quad \vec{c} = (c, 0). \tag{4.3}$$

We are now ready to reap the first fruits of our labor. Let f be a rational function, and replace all occurrences of the variable x with $\vec{x}$, each constant c_i with $\vec{c}_i$, and all arithmetic operations with their counterparts from (4.2). This produces the new function $\vec{f}$ which, upon evaluation at $(x_0, 1)$, produces the ordered pair $(f(x_0), f'(x_0))$.

Example 4.1.1 *Let* $f(x) = \frac{(x+1)(x-2)}{x+3}$. *We wish to compute the values of* $f(3)$ *and* $f'(3)$. *It is easy to see that* $f(3) = 2/3$. *The value of* $f'(3)$, *however, is not immediate. Applying the techniques of differentiation arithmetic, we define*

$$\vec{f}(\vec{x}) = \frac{(\vec{x} + \vec{1})(\vec{x} - \vec{2})}{\vec{x} + \vec{3}} = \frac{((x, 1) + (1, 0)) \times ((x, 1) - (2, 0))}{(x, 1) + (3, 0)}.$$

Inserting the value $\vec{x} = (3, 1)$ *into* $\vec{f}$ *produces*

$$\vec{f}(3, 1) = \frac{((3, 1) + (1, 0)) \times ((3, 1) - (2, 0))}{(3, 1) + (3, 0)}$$

$$= \frac{(4, 1) \times (1, 1)}{(6, 1)} = \frac{(4, 5)}{(6, 1)} = \left(\frac{2}{3}, \frac{13}{18}\right).$$

From this calculation it follows that $f(3) = 2/3$ *(which we already knew) and* $f'(3) = 13/18$. *Note that we never used the expression for* f'.

If we use the different (but equivalent) representation $f(x) = x - \frac{4x+2}{x+3}$, *the same procedure as above yields*

$$\vec{f}(3, 1) = (3, 1) - \frac{(4, 0) \times (3, 1) + (2, 0)}{(3, 1) + (3, 0)}$$

$$= (3, 1) - \frac{(12, 4) + (2, 0)}{(6, 1)} = (3, 1) - \frac{(14, 4)}{(6, 1)}$$

$$= (3, 1) - \left(\frac{7}{3}, \frac{5}{18}\right) = \left(\frac{2}{3}, \frac{13}{18}\right).$$

Thus we arrive at the same result by a completely different route.

Implementing the class constructor is straightforward in MATLAB.

```
01 function ad = autodiff(val, der)
02 % A naive autodiff constructor.
03 ad.val = val;
04 if nargin == 1
05     der = 0.0;
06 end
07 if strcmp(der,'variable')
08     der = 1.0;
09 end
10 ad.der = der;
11 ad = class(ad, 'autodiff');
```

Here, lines 04–06 automatically cast a real number c into an automatic differentiation (AD) type constant $\vec{c} = (c, 0)$, whereas lines 07–09 manually cast a real number x into an AD-type variable $\vec{x} = (x, 1)$.

The display of autodiff objects is handled via display.m:

```
01 function display(ad)
02 % A simple output formatter for the autodiff class.
03 disp([inputname(1), ' = ']);
04 fprintf('   (%17.17f, %17.17f)\n', ad.val, ad.der);
```

We can now input/output autodiff objects within the MATLAB environment:

```
>> a = autodiff(3), b = autodiff(2, 'variable')
a =
   (3.00000000000000000, 0.00000000000000000)
b =
   (2.00000000000000000, 1.00000000000000000)
```

Arithmetic is easy to implement. Here is multiplication according to (4.2):

```
01 function result = mtimes(a, b)
02 % Overloading the '*' operator.
03 [a, b] = cast(a, b);
04 val = a.val*b.val;
05 der = a.val*b.der + a.der*b.val;
06 result = autodiff(val, der);
```

Exercise 4.1. *Write a complete program module that implements the data type* autodiff *defined by the rules (4.2) and (4.3). Use the module to compute the derivatives of more complicated rational functions than the ones used in Examples 4.1.1 and 4.1.2*

Exercise 4.2. *It is straightforward to extend the forward difference* $\Delta_h^+(f, x_0)$ *appearing in (4.1) to the interval version*

$$\Delta_h^+(F, \mathbf{x}_0) \stackrel{\text{def}}{=} \frac{F(\mathbf{x}_0 + h) - F(\mathbf{x}_0)}{h}.$$

Explain why this interval function may not enclose the derivative $f'(x_0)$.

We now want to extend these techniques to elementary functions. This is achieved by implementing the chain rule $(g \circ u)'(x) = u'(x)(g' \circ u)(x)$ in the second component of the ordered pairs. We thus define the rule

$$\vec{g}(\vec{u}) = \vec{g}\big((u, u')\big) = \big(g(u), u'g'(u)\big). \tag{4.4}$$

To give some examples, we define extensions of the following standard functions:

$$
\begin{aligned}
\sin \vec{u} &= \sin (u, u') &&= (\sin u, u' \cos u) \\
\cos \vec{u} &= \cos (u, u') &&= (\cos u, -u' \sin u) \\
e^{\vec{u}} &= e^{(u, u')} &&= (e^u, u'e^u) \\
\log \vec{u} &= \log (u, u') &&= (\log u, u'/u) && (u > 0) \\
\vec{u}^{\alpha} &= (u, u')^{\alpha} &&= (u^{\alpha}, u'\alpha u^{\alpha-1}) && (\alpha \neq 0) \\
|\vec{u}| &= |(u, u')| &&= (|u|, u'\text{sign}(u)) && (u \neq 0).
\end{aligned}
\tag{4.5}
$$

This list can be complemented by more functions, such as $\tan x$, $\arcsin x$, and so forth. We can then automatically generate the first derivative of any elementary function by combining the rules (4.2), (4.4), and (4.5).

Example 4.1.2 *Let $f(x) = (1 + x + e^x) \sin x$, and suppose we want to compute $f'(0)$. As in Example 4.1.1, we define the extended function*

$$\vec{f}(\vec{x}) = (\vec{1} + \vec{x} + e^{\vec{x}}) \sin \vec{x},$$

and evaluate it at $\vec{x} = (0, 1)$. This gives

$$
\begin{aligned}
\vec{f}(0, 1) &= \big((1, 0) + (0, 1) + e^{(0,1)}\big) \sin (0, 1) \\
&= \big((1, 1) + (e^0, e^0)\big)(\sin 0, \cos 0) = (2, 2)(0, 1) = (0, 2).
\end{aligned}
$$

From this simple calculation, it follows that $f(0) = 0$ and $f'(0) = 2$.

Here is the MATLAB implementation for the logarithm:

```
01 function result = log(a)
02 % Overloading the 'log' operator.
03 if (a.val <= 0.0)
04    error('log undefined for non-positive arguments.');
05 end
06 val = log(a.val);
07 der = a.der/a.val;
08 result = autodiff(val, der);
```

Exercise 4.3. *Extend your module using (4.5) and (4.4) so it handles elementary functions, too.*

Once we have implemented enough standard functions, we can put the module to work. Here is a simple MATLAB program that returns the derivative of a general function f at a given point x_0:

```
01 function dx = computeDerivative(fcnName, x0)
02 f  = inline(fcnName);
03 x  = autodiff(x0, 'variable');
04 dx = getDer(f(x));
```

The subfunction `getDer` simply returns the second component (i.e., the derivative value) of the `autodiff` class. A typical usage of this small program is:

```
>> dfx = computeDerivative('(1 + x + exp(x))*sin(x)', 0)
dfx =
    2
>> dfx = computeDerivative('exp(sin(exp(cos(x)+2*power(x,5))))', 1)
dfx =
    129.6681309181679
```

Note that it is possible to combine interval arithmetic with differentiation arithmetic. We then compute with ordered pairs of intervals $\vec{u} = ([u], [u'])$. The result of such a computation is a pair of intervals enclosing the range of the function and its derivative, respectively.

Example 4.1.3 *Let* $f(x) = 1 + \sin(2x)$, *and suppose that we want to compute the enclosure* $F'([0, \frac{\pi}{4}])$. *As in the previous examples, we define the extended function*

$$\vec{F}(\vec{x}) = \vec{1} + \sin(2\vec{x}),$$

except that we now evaluate $\vec{F}$ *at the interval pair* $\vec{x} = ([0, \frac{\pi}{4}], 1)$. *This gives*

$$\vec{F}([0, \tfrac{\pi}{4}], 1) = (1, 0) + \sin\big((2, 0)([0, \tfrac{\pi}{4}], 1)\big) = (1, 0) + \sin([0, \tfrac{\pi}{2}], 2)$$

$$= (1, 0) + (\sin[0, \tfrac{\pi}{2}], 2\cos[0, \tfrac{\pi}{2}]) = (1, 0) + ([0, 1], 2[0, 1])$$

$$= (1, 0) + ([0, 1], [0, 2]) = ([1, 2], [0, 2]).$$

From this calculation, it follows that $F([0, \frac{\pi}{4}]) = [1, 2]$ *and* $F'([0, \frac{\pi}{4}]) = [0, 2]$. *It just happens (due to monotonicity) that both enclosures are sharp:* $R(f; [0, \frac{\pi}{4}]) = [1, 2]$ *and* $R(f'; [0, \frac{\pi}{4}]) = [0, 2]$.

Exercise 4.4. *Write a short program that links your interval module with the automatic differentiation module just created. Given an elementary function* f *and an interval* x, *the program should return enclosures of* $R(f; x)$ *and* $R(f'; x)$.

4.2 HIGHER-ORDER DERIVATIVES

It should come as no surprise that the techniques presented in the previous section can be generalized to produce derivatives of higher order than one. As an example, for second-order derivatives, we must compute with triples of numbers $\vec{u} = (u, u', u'')$, and the arithmetic rules corresponding to (4.2) generalize to

$$\vec{u} + \vec{v} = (u + v, u' + v', u'' + v'')$$
$$\vec{u} - \vec{v} = (u - v, u' - v', u'' - v'') \tag{4.6}$$
$$\vec{u} \times \vec{v} = (uv, uv' + u'v, uv'' + 2u'v' + u''v)$$
$$\vec{u} \div \vec{v} = (u/v, (u' - (u/v)v')/v, (u'' - 2(u/v)'v' - (u/v)v'')/v),$$

where we, as before, demand that $v \neq 0$ when dividing. Extending the second-order rules to the standard functions is straightforward but tedious.[1]

---

[1] For a complete C++ implementation of second-order automatic differentiation, see [CXSC].

A more effective (and perhaps less error-prone) approach to high-order automatic differentiation is obtained through the calculus of Taylor series (see, e.g., [Mo66], [Mo79], [Ab88], or [Ab98]). Given a real-valued function $f \in C^\infty$, its Taylor expansion at x_0 is

$$f(x) = f_0 + f_1(x - x_0) + \cdots + f_k(x - x_0)^k + \cdots,$$

where the Taylor coefficients are given by $f_k = f_k(x_0) = f^{(k)}(x_0)/k!$. Here, we are using the common notation for derivatives: $f^{(k)} = \frac{d^k f}{dx^k}$. Now, since the Taylor coefficients are just rescaled derivatives, it follows that computing the derivatives of a function is equivalent to computing its Taylor series.

It is clear that adding the Taylor series of two functions f and g produces a new Taylor series corresponding the function $f + g$, and similarly for subtraction. The Taylor coefficients of the product and quotient of two functions are slightly more complicated to compute. Before deriving these, we summarize the rules for Taylor arithmetic:

$$(f + g)_k = f_k + g_k$$
$$(f - g)_k = f_k - g_k$$
$$(f \times g)_k = \sum_{i=0}^{k} f_i g_{k-i} \qquad (4.7)$$
$$(f \div g)_k = \frac{1}{g_0} \left(f_k - \sum_{i=0}^{k-1} (f \div g)_i g_{k-i} \right).$$

Note that when dividing, we must[2] have $g_0 \neq 0$, which corresponds to the restriction $v \neq 0$ in (4.6) and (4.2).

To prove the rule for multiplication, we simply write

$$\sum_{k=0}^{\infty} f_k(x - x_0)^k \sum_{k=0}^{\infty} g_k(x - x_0)^k = \sum_{k=0}^{\infty} (f \times g)_k(x - x_0)^k$$

from which the coefficient $(f \times g)_k$ is obtained by gathering all powers $(x - x_0)^k$ from the left-hand side:

$$\sum_{i=0}^{k} f_i(x - x_0)^i g_{k-i}(x - x_0)^{k-i} = (f \times g)_k(x - x_0)^k.$$

The rule for division is obtained in a similar fashion: by definition, we have

$$\sum_{k=0}^{\infty} f_k(x - x_0)^k \bigg/ \sum_{k=0}^{\infty} g_k(x - x_0)^k = \sum_{k=0}^{\infty} (f \div g)_k(x - x_0)^k.$$

Multiplying both sides with the Taylor series for g produces

$$\sum_{k=0}^{\infty} f_k(x - x_0)^k = \sum_{k=0}^{\infty} (f \div g)_k(x - x_0)^k \sum_{k=0}^{\infty} g_k(x - x_0)^k,$$

---

[2] Actually, if $g_i = f_i = 0$ for $i = 0, \ldots, m - 1$ and $g_m \neq 0$, then the division can be carried out according to l'Hopital's rule.

and by the rule for multiplication, we have

$$f_k = \sum_{i=0}^{k} (f \div g)_i g_{k-i} = \sum_{i=0}^{k-1} (f \div g)_i g_{k-i} + (f \div g)_k g_0.$$

Solving for $(f \div g)_k$ produces the desired result.

In order to be able to compute with this new arithmetic, we need to know how constants and the independent variable x are treated. Seen as functions, these have particularly simple Taylor expansions:

$$
\begin{aligned}
x &= x_0 & +1 \cdot (x - x_0) + 0 \cdot (x - x_0)^2 + \cdots + 0 \cdot (x - x_0)^k + \cdots, \\
c &= c & +0 \cdot (x - x_0) + 0 \cdot (x - x_0)^2 + \cdots + 0 \cdot (x - x_0)^k + \cdots.
\end{aligned}
\tag{4.8}
$$

We now illustrate how simple it is to implement a `taylor` class constructor in MATLAB.

```
01 function ts = taylor(a, N, str)
02 % A naive taylor constructor.
03 if nargin == 1
04     if isa(a,'taylor')
05         ts = a;
06     else
07         ts.coeff = a;
08     end
09 elseif nargin == 3
10     ts.coeff = zeros(1,N);
11     if strcmp(str,'variable')
12         ts.coeff(1) = a;   ts.coeff(2) = 1;
13     elseif strcmp(str,'constant');
14         ts.coeff(1) = a;   ts.coeff(2) = 0;
15     end
16 end
17  ts = class(ts, 'taylor');
```

In this implementation we have chosen to make use of *explicit* casting, that is, on lines 11–14, we decide how a scalar value is converted into a `taylor` object. As before, we also need to implement the way to display the class objects:

```
01 function display(ts)
02 % A simple output formatter for the taylor class.
03 disp([inputname(1), ' = ']);
04 fprintf('[')
05 for i=1:length(ts.coeff)-1
06     fprintf('%17.17f, ', ts.coeff(i));
07 end
08 fprintf('%17.17f]\n', ts.coeff(end));
```

We can now input/output `taylor` objects within the MATLAB environment:

```
>> x = taylor(1.5, 3, 'variable'), c = taylor(pi, 2, 'constant')
x =
  [1.50000000000000000, 1.00000000000000000, 0.00000000000000000]
c =
  [3.14159265358979312, 0.00000000000000000]
```

Following (4.7), we can write an implementation for division

```
01 function result = mrdivide(a, b)
02 % Overloading the '/' operator.
03 [a, b] = cast(a, b);
04 if (b.coeff(1) == 0.0)
05     error('Denominator is zero.');
06 else
07     N = length(a.coeff);
08     coeff = zeros(1,N);
09     coeff(1) = a.coeff(1)/b.coeff(1);
10     for k=1:N-1
11         sum = a.coeff(k + 1);
12         for i=0:k-1
13             sum = sum - coeff(i+1)*b.coeff(k-i+1);
14         end
15         coeff(k+1) = sum/b.coeff(1);
16     end
17     result = taylor(coeff);
18 end
```

Here, the `cast` function makes sure that both inputs are of `taylor` type and have the same length.[3]

Exercise 4.5. *Complete the program module that implements the data type* `taylor` *defined by the rules (4.7) and (4.8). Use the module to compute high-order derivatives of the rational function appearing in Example 4.1.1 for several expansion points.*

Exercise 4.6. *Extend the rule for division of (4.7), incorporating the case* $g_i = f_i = 0$ *for* $i = 0, \ldots, m - 1$ *and* $g_m \neq 0$. *Explain how this extension corresponds to l'Hopital's rule.*

Exercise 4.7. *Write down the formal expression for* $f \times f$ *using the rule for multiplication of (4.7). Using the appearing symmetry, find a more efficient formula for computing the square* f^2 *of a function* f.

4.2.1 Derivatives of Standard Functions

Extending these ideas to the standard functions is not difficult. Let us start with exponentiation. Given a function g whose Taylor series is known, how do we compute the Taylor series for e^g? Let us formally write

$$g(x) = \sum_{k=0}^{\infty} g_k(x - x_0)^k \quad \text{and} \quad e^{g(x)} = \sum_{k=0}^{\infty} (e^g)_k(x - x_0)^k,$$

and use the fact that

$$\frac{d}{dx}e^{g(x)} = g'(x)e^{g(x)}. \tag{4.9}$$

---

[3]We append zero coefficients to the shorter expansion. Note, however, that this is not always the appropriate design choice.

Inserting the formal expressions for $g'(x)$ and $e^{g(x)}$ into (4.9) produces

$$\sum_{k=1}^{\infty} k(e^g)_k (x - x_0)^{k-1} = \sum_{k=1}^{\infty} kg_k (x - x_0)^{k-1} \sum_{k=0}^{\infty} (e^g)_k (x - x_0)^k,$$

which, after multiplying both sides with $(x - x_0)$, becomes

$$\sum_{k=1}^{\infty} k(e^g)_k (x - x_0)^k = \sum_{k=1}^{\infty} kg_k (x - x_0)^k \sum_{k=0}^{\infty} (e^g)_k (x - x_0)^k.$$

Using the rule for products from (4.7) then yields

$$k(e^g)_k = \sum_{i=1}^{k} i g_i (e^g)_{k-i} \qquad (k > 0).$$

Since we know that the constant term is given by $(e^g)_0 = e^{g_0}$, we arrive at

$$(e^g)_k = \begin{cases} e^{g_0} & \text{if } k = 0, \\ \dfrac{1}{k} \displaystyle\sum_{i=1}^{k} i g_i (e^g)_{k-i} & \text{if } k > 0. \end{cases} \tag{4.10}$$

The natural logarithm of a function $\ln g(x)$ can be obtained in a similar fashion. Using the relation

$$\frac{d}{dx} \ln g(x) = \frac{g'(x)}{g(x)},$$

we find that

$$\sum_{k=1}^{\infty} k(\ln g)_k (x - x_0)^{k-1} = \sum_{k=1}^{\infty} kg_k (x - x_0)^{k-1} \Bigg/ \sum_{k=0}^{\infty} g_k (x - x_0)^k.$$

Multiplying both sides with $(x - x_0)$ and rearranging produces

$$\sum_{k=1}^{\infty} k(\ln g)_k (x - x_0)^k \sum_{k=0}^{\infty} g_k (x - x_0)^k = \sum_{k=1}^{\infty} kg_k (x - x_0)^k.$$

Using the rule for multiplication from (4.7), we have

$$\sum_{i=1}^{k} i(\ln g)_i g_{k-i} = kg_k,$$

and solving for $(\ln g)_k$ yields[4]

$$(\ln g)_k = \frac{1}{g_0} \left(g_k - \frac{1}{k} \sum_{i=1}^{k-1} i(\ln g)_i g_{k-i} \right) \qquad (k > 0).$$

---

[4]When $k = 1$, the sum appearing in the formula for $(\ln g)_k$ is empty and evaluates to zero.

Since we know that the constant term is given by $(\ln g)_0 = \ln g_0$, we arrive at

$$(\ln g)_k = \begin{cases} \ln g_0 & \text{if } k = 0, \\ \frac{1}{g_0}\left(g_k - \frac{1}{k}\sum_{i=1}^{k-1} i(\ln g)_i g_{k-i}\right) & \text{if } k > 0. \end{cases} \tag{4.11}$$

We can now define exponentiation through the equality

$$g(x)^{f(x)} = e^{f(x)\ln g(x)}. \tag{4.12}$$

If, however, the exponent is a constant, $f(x) = a$, we can find a more effective means for computing $g(x)^a$. We will use the fact that we have

$$\frac{d}{dx}g(x)^a = ag'(x)g(x)^{a-1} \quad \text{or} \quad g(x)\frac{d}{dx}g(x)^a = ag'(x)g(x)^a. \tag{4.13}$$

The Taylor series representation for the second part of (4.13) is

$$\sum_{k=0}^{\infty} g_k(x-x_0)^k \sum_{k=1}^{\infty} k(g^a)_k(x-x_0)^{k-1} = a\sum_{k=1}^{\infty} kg_k(x-x_0)^{k-1}$$

$$\times \sum_{k=0}^{\infty} (g^a)_k(x-x_0)^k.$$

Multiplying both sides with $(x-x_0)$ and using the rule for multiplication from (4.7) gives

$$\sum_{i=0}^{k} g_i(k-i)(g^a)_{k-i} = a\sum_{i=1}^{k} ig_i(g^a)_{k-i}.$$

Extracting the first term of the left-hand side produces

$$g_0 k(g^a)_k = \sum_{i=1}^{k}\left(aig_i(g^a)_{k-i} - g_i(k-i)(g^a)_{k-i}\right)$$

$$= \sum_{i=1}^{k}\left((a+1)i - k\right)g_i(g^a)_{k-i}.$$

Summarizing, we arrive at:

$$(g^a)_k = \begin{cases} g_0^a & \text{if } k = 0, \\ \frac{1}{g_0}\sum_{i=1}^{k}\left(\frac{(a+1)i}{k} - 1\right)g_i(g^a)_{k-i} & \text{if } k > 0. \end{cases} \tag{4.14}$$

Exercise 4.8. *Unfortunately, neither formula (4.12) nor (4.14) is suitable for the situation when $g_0 = 0$ and $a > 0$, despite the fact that $g(x)^a$ may be well-defined. Try to derive a suitable recursive formula for this case.*

The Taylor coefficients of $\sin g(x)$ and $\cos g(x)$ must be computed in parallel. Following the usual procedure of differentiating, and multiplying both sides by

$(x - x_0)$, we end up with the formulas:

$$(\sin g)_k = \begin{cases} \sin g_0 & \text{if } k = 0, \\ \dfrac{1}{k}\displaystyle\sum_{i=1}^{k} i g_i (\cos g)_{k-i} & \text{if } k > 0. \end{cases}$$

(4.15)

$$(\cos g)_k = \begin{cases} \cos g_0 & \text{if } k = 0, \\ -\dfrac{1}{k}\displaystyle\sum_{i=1}^{k} i g_i (\sin g)_{k-i} & \text{if } k > 0. \end{cases}$$

By using the rule for division from (4.7), it is straightforward to obtain the Taylor coefficients of $\tan g(x)$.

$$(\tan g)_k = \begin{cases} \tan g_0 & \text{if } k = 0, \\ \dfrac{1}{\cos^2 g_0}\left(g_k - \dfrac{1}{k}\displaystyle\sum_{i=1}^{k-1} i (\tan g)_i (\cos^2 g)_{k-i}\right) & \text{if } k > 0. \end{cases}$$

(4.16)

For the inverse trigonometric functions, we use the well-known rules of differentiation

$$\frac{d}{dx} \arcsin g(x) = \frac{g'(x)}{\sqrt{1 - (g(x))^2}}$$

$$\frac{d}{dx} \arccos g(x) = \frac{-g'(x)}{\sqrt{1 - (g(x))^2}}$$

$$\frac{d}{dx} \arctan g(x) = \frac{g'(x)}{1 + (g(x))^2},$$

which yield the recursive formulas:

$$(\arcsin g)_k = \begin{cases} \arcsin g_0 & \text{if } k = 0, \\ \dfrac{1}{\sqrt{1 - (g_0)^2}}\left(g_k - \dfrac{1}{k}\displaystyle\sum_{i=1}^{k-1} i (\arcsin g)_i (\sqrt{1 - g^2})_{k-i}\right) & \text{if } k > 0, \end{cases}$$

$$(\arccos g)_k = \begin{cases} \arccos g_0 & \text{if } k = 0, \\ \dfrac{-1}{\sqrt{1 - (g_0)^2}}\left(g_k + \dfrac{1}{k}\displaystyle\sum_{i=1}^{k-1} i (\arccos g)_i (\sqrt{1 - g^2})_{k-i}\right) & \text{if } k > 0, \end{cases}$$

$$(\arctan g)_k = \begin{cases} \arctan g_0 & \text{if } k = 0, \\ \dfrac{1}{1 + (g_0)^2}\left(g_k - \dfrac{1}{k}\displaystyle\sum_{i=1}^{k-1} i (\arctan g)_i (1 + g^2)_{k-i}\right) & \text{if } k > 0. \end{cases}$$

Once we have implemented sufficiently many standard functions for our taylor class, we can use the following MATLAB program for arbitrary order differentiation:

```
01 function dx = computeDerivative(fcnName, x0, order)
02 f   = inline(fcnName);
03 x   = taylor(x0, order+1, 'variable');
04 dx = getDer(f(x), order);
```

Here, `getDer` converts a Taylor coefficient into a derivative by multiplying it by the proper factorial. A typical usage is the following:

```
>> df40 = computeDerivative('exp(sin(exp(cos(x) + 2*x^5)))', -2, 40)
df40 =
         1.4961e+53
```

Of course, for such high derivatives, the effects of rounding errors may have rendered the result incorrect. Fortunately, it is not hard to change the underlying scalars to intervals. This takes the whole machinery of the preceding chapters into account, guaranteeing a validated result.[5]

Exercise 4.9. *Extend your implementation of the data type* `taylor` *to incorporate the recursive rules for the standard functions presented above. Use the module to compute the quantity* $f^{(4)}(1)$, *where* $f(x) = (5 + \cos^2 3x)e^{x+\sin 7x}$.

4.3 HIGHER-ORDER ENCLOSURES

If f is n times continuously differentiable on a domain $\mathbf{x}$, then we can expand f in its Taylor series around any point $x_0 \in \mathbf{x}$:

$$f(x) = f(x_0) + f'(x_0)(x - x_0) + \cdots + \frac{f^{(n-1)}(x_0)}{(n-1)!}(x - x_0)^{n-1}$$
$$+ \frac{f^{(n)}(\zeta)}{n!}(x - x_0)^n$$
$$= f_0 + f_1(x - x_0) + \cdots + f_{n-1}(x - x_0)^{n-1} + \frac{f^{(n)}(\zeta)}{n!}(x - x_0)^n \quad (4.17)$$

where $x \in \mathbf{x}$ and ζ is somewhere between x and x_0. By combining automatic differentiation and interval arithmetic, we obtain enclosures of the Taylor coefficients, which we denote $F_0, F_1, \ldots, F_n$. In particular, we have $\frac{f^{(n)}(\zeta)}{n!} \in F_n$ for all $\zeta \in \mathbf{x}$. Therefore we can bound the remainder term in (4.17) to obtain the enclosure

$$f(x) \in f_0 + f_1(x - x_0) + \cdots + f_{n-1}(x - x_0)^{n-1} + F_n(\mathbf{x} - x_0)^n, \quad (4.18)$$

valid for all $x_0, x \in \mathbf{x}$.

Note that if we select $x_0 = \text{mid}(\mathbf{x})$, and let $r = \text{rad}(\mathbf{x})$, then the remainder term is enclosed by the interval $F_n r^n[-1, 1]$ if n is odd, and $F_n r^n[0, 1]$ if n is even. In both cases, the interval widths scale like r^n.

---

[5]In this example, an interval-based automatic differentiation implementation yields the enclosure
`df40` $\in$ 1.49[5513, 6706].

To emphasize the splitting of f into a real-valued polynomial part and an interval-valued remainder term, we introduce the special notations

$$P_{f,x_0}^{n-1}(h) \stackrel{\text{def}}{=} f_0 + f_1 h + \cdots + f_{n-1} h^{n-1},$$

$$I_{f,x}^n(h) \stackrel{\text{def}}{=} F^{(n)}(x)[-h, h]^n.$$

We then have the following inclusion:

$$f(x_0 + h) \in P_{f,x_0}^{n-1}(h) + I_{f,x}^n(h), \tag{4.19}$$

which also provides an enclosure of the range of f over $x = x_0 + [-r, r]$:

$$R(f; x) \subseteq R\left(P_{f,x_0}^{n-1}; [-r, r]\right) + I_{f,x}^n(r).$$

The idea of separating a function into a polynomial approximation and an interval enclosure of the remainder term, as in (4.19), is really not new: we have already seen this in the centered forms, where the polynomial part has degree zero.[6]

4.4 COMPUTER LAB III

Problem 1. Implement the first-order automatic differentiation rules (4.2), (4.3), and (4.5). The underlying number field should be of type `double`. Use your implementation to compute the couple $(f(x_0), f'(x_0))$, where $x_0 = 1$ and f is given in Problem 2 of Computer Lab II.

Problem 2. Repeat Problem 1, but this time the underlying number field should be of type `interval`. Now use your implementation to compute the eleven couples $(f(x_k), f'(x_k))$ for $k = 0, \ldots, 10$, where $x_k = 1 + 2^{-k}[-1, 1]$.

Problem 3. Write a small set of routines supporting Taylor series arithmetic, as defined by equations (4.7) and (4.8). The underlying number field should be of type `double`. Use your implementation to compute $g^{(k)}(1)$ for $k = 0, \ldots, 5$, where

$$g(x) = \frac{7x - (x + 1)^2}{3x - 2}.$$

(Recall that $g^{(k)}(x) = \frac{d^k g}{dx^k}(x)$.)

Problem 4. Using the formulas from Section 4.2.1, extend your work from Problem 3 by implementing Taylor series arithmetic versions of the standard functions e^x, $\log x$, x^n ($n \in \mathbb{Z}$), x^α ($\alpha \in \mathbb{R} \setminus \mathbb{Z}$), $\sin x$, $\cos x$, and $\tan x$. Finally, compute $f^{(k)}(1)$ for $k = 0, \ldots, 5$ where f is given in Problem 2 of Computer Lab II.

---

[6]A more structured development of so-called Taylor models was introduced by Berz and Makino (see, e.g., [Be97] and [BM98]). They take the whole concept further by defining a Taylor-model arithmetic, which allows them to compute with objects of the form similar to (4.19) as intrinsic quantities. The main difference between these and what we have just done is that they provide for the propagation of the involved error terms. This includes the absorption of higher-order (or otherwise unwanted) coefficients into the error terms. Pursuing this interesting direction of research would, however, take us too far afield from our present path.

Chapter Five

Interval Analysis in Action

IN THIS CHAPTER, we will illustrate how interval analysis can be successfully applied to various areas of mathematics.

5.1 ZERO-FINDING METHODS

We will start by presenting methods for locating all zeros of a function f over a given domain $\mathbf{x}$. If the function is merely continuous, a bisection-type method is usually a good choice, whereas if f is differentiable, a Newton-type method may be preferable. We will begin by describing the simpler bisection-type method.

5.1.1 Divide and Conquer

Let $f : I \to \mathbb{R}$ be a continuous, elementary function. By Theorem 3.8 we know that if $\mathbf{x} \subseteq I$ and $F(\mathbf{x})$ exists, then $R(f; \mathbf{x}) \subseteq F(\mathbf{x})$. As pointed out earlier, this is equivalent to the statement "$y \notin F(\mathbf{x}) \Rightarrow y \notin R(f; \mathbf{x})$." Our approach to finding all zeros of f will be an adaptive bisection scheme, where subsets of the initial domain either are proved *not* to contain any zeros, whereby they are discarded from the search, or are bisected and kept for further study. When all remaining subintervals are smaller than some predefined tolerance, the search is terminated. We are then left with a (possibly empty) collection of intervals whose union covers all possible zeros of f within $\mathbf{x}$. A simple implementation of this search is presented in Listing 5.1. The implementation uses the PROFIL/BIAS interval package (see [PrBi]).

Upon the termination of this program, all elements of the output stream will be possible candidates for zero enclosures for f. That is, an interval printed from the program *may* contain a zero of f. What is *absolutely sure* is that no zero of f can exist outside the union of the printed elements.

Example 5.1.1 *Consider the function $f(x) = \sin x(x - \cos x)$ over the domain* $[-10, 10]$, *illustrated in Figure 5.1(a). The function clearly has eight zeros in the domain:* $\{\pm 3\pi, \pm 2\pi, \pm \pi, 0, x^*\}$, *where x^* is the unique (positive) zero of* $x - \cos x = 0$. *Executing Listing 5.1 with tolerance 0.001 produces the nine intervals listed below. Note, however, that intervals 4 and 5 are adjacent. This always happens when a zero is located exactly at a bisection point of the domain.*

Listing 5.1.   An implementation of a recursive interval bisection scheme using the
PROFILE/BIAS package [PrBi]

```
1  #include <iostream>
2  #include "Interval.h"
3  #include "Functions.h"
4  using namespace std;
5  typedef INTERVAL (*pfcn)(INTERVAL);
6
7  void bisect(pfcn f, INTERVAL X, double Tol) {
8    if ( 0.0 <= f(X) )
9      if ( Diam(X) < Tol )
10        cout << X << endl;
11      else {
12        bisect(f, INTERVAL(Inf(X), Mid(X)), Tol);
13        bisect(f, INTERVAL(Mid(X), Sup(X)), Tol);
14      }
15  }
16
17  INTERVAL function(INTERVAL x) {
18    return sin(x)*(x - cos(x));
19  }
20
21  int main(int argc, char * argv[])
22  {
23    INTERVAL   X(atof(argv[1]), atof(argv[2]));
24    double   Tol(atof(argv[3]));
25
26    bisect(function, X, Tol);
27    return 0;
28  }
```

```
Domain          :  [-10,10]
Tolerance       :  0.001
Function calls: 227
Root list       :
                  1:  [-9.42505,-9.42444]   6:  [+0.73853,+0.73914]
                  2:  [-6.28357,-6.28296]   7:  [+3.14148,+3.14209]
                  3:  [-3.14209,-3.14148]   8:  [+6.28296,+6.28357]
                  4:  [-0.00061,+0.00000]   9:  [+9.42444,+9.42505]
                  5:  [+0.00000,+0.00061]
```

Exercise 5.1. *Write your own bisection routines: one using the interval techniques
described in this section, and one using the standard floating point approach. Evaluate the two routines by comparing the number of function evaluations needed to
locate all zeros within some tolerance. [Co77] provides some interesting ideas.*

Note that if we use a centered form when computing the enclosure of $R(f; \tilde{x})$,
we can use the information provided from the derivative enclosure $F'(\tilde{x})$. Recall
that if $0 \notin F'(\tilde{x})$, then f is monotone over $\tilde{x}$. If this is the case, then f either has
a unique zero in $\tilde{x}$ or has no zeros in $\tilde{x}$ at all. The former happens when f has
opposite signs at the endpoints of $\tilde{x}$, the latter when the signs are equal.

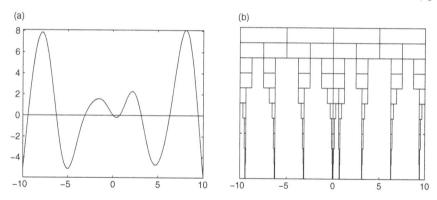

Figure 5.1 (a) The function $f(x) = \sin x (x - \cos x)$ over the domain $[-10, 10]$; (b) Increasingly tight zero enclosures.

Exercise 5.2. *Rewrite Listing 5.1, incorporating checks for monotonicity.*

5.1.2 Newton's Method

Assuming now that the function is differentiable, we will continue to illustrate the power of interval arithmetic in zero finding. The basic tool for finding zeros of non-linear equations is Newton's method, which we will begin by describing in the real-valued setting.

Let $f : \mathbf{x} \rightarrow \mathbb{R}$ be a continuously differentiable function, and suppose that $x^* \in \mathbf{x}$ is a zero of f, that is, $f(x^*) = 0$. Given an initial guess $x_0 \in \mathbf{x}$, our strategy is to compute the intersection between the tangent at $f(x_0)$ and the x-axis. The tangent is given by the linear equation

$$t(x) = f(x_0) + f'(x_0)(x - x_0),$$

so the point of intersection is easily computed as

$$x_1 = x_0 - \frac{f(x_0)}{f'(x_0)}.$$

This expression is well-defined, provided that $f'(x_0) \neq 0$. Repeating this process gives the following sequence (see Figure 5.2):

$$x_{k+1} = x_k - \frac{f(x_k)}{f'(x_k)} \qquad k = 0, 1, 2, \ldots.$$

THEOREM 5.3. *Assume that $f : \mathbf{x} \rightarrow \mathbb{R}$ is twice continuously differentiable with $f'(x) \neq 0$ for all $x \in \mathbf{x}$ and that f has a unique, simple zero x^* within $\mathbf{x}$. Then, if x_0 is chosen sufficiently close to x^*, the sequence $\{x_k\}_{k=0}^{\infty}$ converges quadratically fast toward x^*, that is, there exists a constant C such that*

$$\lim_{k \to \infty} x_k = x^* \qquad and \qquad |x_{k+1} - x^*| \leq C |x_k - x^*|^2.$$

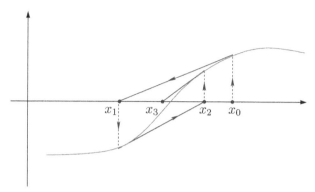

Figure 5.2 The first few iterates of Newton's method.

Proof. Let ε_k denote the error in the iterate x_k, that is, $\varepsilon_k = x_k - x^*$. Expanding f in a Taylor series centered at x_k gives

$$0 = f(x^*) = f(x_k) + f'(x_k)(x^* - x_k) + \tfrac{1}{2}f''(\zeta)(x^* - x_k)^2,$$

where ζ is between x^* and x_k. Dividing by $f'(x_k)$ then gives

$$\frac{f(x_k)}{f'(x_k)} + x^* - x_k = x^* - x_{k+1} = \frac{-\tfrac{1}{2}f''(\zeta)(x^* - x_k)^2}{f'(x_k)},$$

which means that

$$|\varepsilon_{k+1}| = \left| \frac{f''(\zeta)}{2f'(x_k)} \right| \varepsilon_k^2 \leq \sup\left\{ \left| \frac{f''(x)}{2f'(y)} \right| : x, y \in \mathbf{x} \right\} \varepsilon_k^2 = C\varepsilon_k^2.$$

We can rewrite the inequality $|\varepsilon_{k+1}| \leq C\varepsilon_k^2$ as $|C\varepsilon_{k+1}| \leq (C\varepsilon_k)^2$, so if $|C\varepsilon_0| < 1$ and $[x^* - \varepsilon_0, x^* + \varepsilon_0] \subseteq \mathbf{x}$, then it follows by induction that $x_k \in \mathbf{x}$ for all k, and that

$$|\varepsilon_k| \leq \frac{1}{C}(C\varepsilon_0)^{2^k}.$$

Therefore, if we choose x_0 close enough to x^* to ensure that $|C\varepsilon_0| = C|x_0 - x^*| < 1$, then it follows that x_k tends to x^* quadratically fast. $\qquad\square$

It should be pointed out that unless the assumptions of Theorem 5.3 are fulfilled, Newton's method may fail to converge. The iterates can form a periodic orbit or wander around in an apparently aimless pattern. This should not be viewed as a rare, degenerate situation but rather as a naturally occurring phenomenon associated to deep mathematical properties of Newton's method.[1]

Using the `autodiff` class developed in Chapter 4, we can write a small MATLAB routine for solving non-linear equations, using Newton's method.

```
01 function y = newtonSearch(fcnName, x, tol)
02 f = inline(fcnName);
```

---

[1] The use of Newton's method to solve cubic polynomials (over $\mathbb{C}$) actually led to the discovery of a simple, deterministic system that exhibits chaotic behavior.

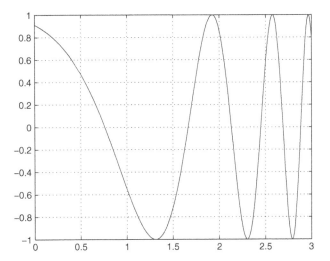

Figure 5.3 The graph of $\sin(e^x + 1)$.

```
03 y = newtonStep(f, x);
04 while (abs(x-y) > tol)
05     x = y;
06     y = newtonStep(f, x);
07 end
08 end
09
10 function Nx = newtonStep(f, x)
11 xx = autodiff(x, 'variable');
12 fx = f(xx);
13 Nx = x-getVal(fx)/getDer(fx);
14 end
```

Note that this function "hides" the automatic differentiation from the user: all input/output is scalar. Here are some sample outputs (see also Figure 5.3):

```
>> x = newtonSearch('sin(exp(x) + 1)', 1, 1e-10)
x =
   0.761549782880894
>> x = newtonSearch('sin(exp(x) + 1)', 0, 1e-10)
x =
   2.131177121086310
```

5.1.3 The Interval Newton Method

Unlike the real-valued Newton method, the interval version always displays very regular behavior. To begin with, we will assume that $f : \mathbf{x} \to \mathbb{R}$ is a continuously differentiable function and that $x^* \in \mathbf{x}$ is a zero of f. We will also assume that an interval extension of f' exists and satisfies $0 \notin F'(\mathbf{x})$. In particular, this implies that $f'(x) \neq 0$ throughout $\mathbf{x}$. Then, for any $x \in \mathbf{x}$, an application of the Mean Value Theorem gives

$$f(x) = f(x^*) + f'(\zeta)(x - x^*)$$

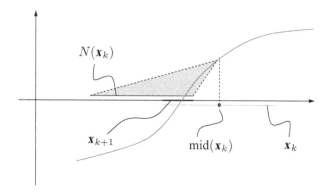

Figure 5.4 One iterate of the interval Newton method.

for some ζ between x and x^*. Since $f(x^*) = 0$ and $f'(\zeta) \neq 0$, we can solve for the zero x^*:

$$x^* = x - \frac{f(x)}{f'(\zeta)} \in x - \frac{f(x)}{F'(\mathbf{x})} \stackrel{\text{def}}{=} N(\mathbf{x}; x).$$

Since we are assuming that $x^* \in \mathbf{x}$, we also have $x^* \in N(\mathbf{x}; x) \cap \mathbf{x}$ for all $x \in \mathbf{x}$. The enclosure corresponding to the selection $x = m = \text{mid}(\mathbf{x})$ is called the *interval Newton operator*:

$$N(\mathbf{x}) \stackrel{\text{def}}{=} N(\mathbf{x}, m) = m - \frac{f(m)}{F'(\mathbf{x})}.$$

Taking $\mathbf{x}_0 = \mathbf{x}$ as our initial enclosure of x^*, we define the sequence of intervals

$$\mathbf{x}_{k+1} = N(\mathbf{x}_k) \cap \mathbf{x}_k, \quad k = 0, 1, 2, \ldots. \tag{5.1}$$

In Figure 5.4 we illustrate the partial construction of this sequence. Note that, by construction, if $x^* \in \mathbf{x}_0$ then $x^* \in \mathbf{x}_k$ for all $k \in \mathbb{N}$. This means that we never lose track of the zero. The sequence of intervals defined by (5.1) has several other nice properties.

THEOREM 5.4. (Interval Newton Method) *Assume that $N(\mathbf{x}_0)$ is well-defined. If $\mathbf{x}_0$ contains a zero x^* of f, then so do all iterates $\mathbf{x}_k$, $k \in \mathbb{N}$. Furthermore, the intervals $\mathbf{x}_k$ form a nested sequence converging to x^*.*

Proof. As mentioned above, the implication $x^* \in \mathbf{x}_0 \Rightarrow x^* \in \mathbf{x}_k$ for all $k \in \mathbb{N}$ follows immediately by induction. Also, by (5.1), it is clear that the intervals $\mathbf{x}_k$ form a nested sequence. We need therefore only prove that the intervals $\mathbf{x}_k$ converge to x^*. This can happen in two ways: the first scenario is that at some stage, there is a thin iterate $\mathbf{x}_k = [x^*, x^*]$, in which case we have $\text{rad}(\mathbf{x}_k) = 0$ and $\mathbf{x}_k = \mathbf{x}_{k+i}$ for all $i \in \mathbb{N}$. This happens exactly when $x^* = \text{mid}(\mathbf{x}_{k-1})$. Let us now consider the second scenario, that is, we assume that $x^* \neq \text{mid}(\mathbf{x}_k)$ for all $k \in \mathbb{N}$. The fact that $N(\mathbf{x}_0)$ is well-defined implies that $0 \notin F'(\mathbf{x}_0)$. Therefore, since $\mathbf{x}_k \subseteq \mathbf{x}_0$, it follows that $0 \notin F'(\mathbf{x}_k)$ for all $k \in \mathbb{N}$. This implies that the correction term

(which now is an interval)

$$\frac{f(\text{mid}(\mathbf{x}_k))}{F'(\mathbf{x}_k)}$$

consists entirely of elements of the same sign. As a consequence, the midpoint of $\mathbf{x}_k$ is not contained in $\mathbf{x}_{k+1}$ (see Figure 5.4). This means that $\text{rad}(\mathbf{x}_{k+1}) < \frac{1}{2}\text{rad}(\mathbf{x}_k)$, and the convergence is proved. □

Thus, the sequence (5.1) converges to x^* at least at a linear rate. Under conditions similar to those of Theorem 5.3, it is possible to show that the convergence rate is quadratic, that is, there exists a constant C such that

$$d(\mathbf{x}_{k+1}, x^*) \leq Cd(\mathbf{x}_k, x^*)^2.$$

A proof of this statement is given in [Mo66].

One of the most useful properties of the interval Newton operator N is that we are provided with a means of detecting when a region does *not* contain a zero of f. As this is a common situation, it is important that we can quickly discard a set on the grounds that it contains no zeros. Another important contribution from the properties of N is a simple, verifiable condition that guarantees the existence of a unique zero within an interval.

THEOREM 5.5. *Assume that $f : \mathbf{x} \to \mathbb{R}$ is twice continuously differentiable, and that $N(\mathbf{x})$ is well-d=efined for some $\mathbf{x} \in \mathbb{IR}$. Then the following statements hold:*

(1) if $N(\mathbf{x}) \cap \mathbf{x} = \emptyset$, then $\mathbf{x}$ contains no zeros of f;
(2) if $N(\mathbf{x}) \subseteq \mathbf{x}$, then $\mathbf{x}$ contains exactly one zero of f.

Proof. We begin by proving part (1). This statement follows as a consequence of Theorem 5.4, since if $\mathbf{x}$ contains a zero x^*, then so does $N(\mathbf{x})$, which means that $x^* \in N(\mathbf{x}) \cap \mathbf{x}$. Therefore, if $N(\mathbf{x}) \cap \mathbf{x} = \emptyset$, then $\mathbf{x}$ cannot contain a zero of f.

To prove part (2), we first recall that $N(\mathbf{x})$ being well-defined implies that f is monotone on $\mathbf{x}$. Therefore, since f is continuous on $\mathbf{x}$, there can be *at most* one zero in $\mathbf{x}$, so we only have to establish the existence of a zero $x^* \in \mathbf{x}$. The inclusion $N(\mathbf{x}) \subseteq \mathbf{x}$ translates to $m - \frac{f(m)}{F'(\mathbf{x})} \subseteq \mathbf{x}$, which implies that

$$m - \frac{f(m)}{f'(\zeta)} \subseteq \mathbf{x} \quad \text{for all} \quad \zeta \in \mathbf{x}.$$

But $m - \frac{f(m)}{f'(\zeta)}$ is the solution to

$$t_\zeta(x) = f(m) + f'(\zeta)(x - m) = 0.$$

Recall that $t_\zeta(x)$ describes the line passing through $(m, f(m))$ and having slope $f'(\zeta)$. Since we are assuming that f' is continuous on the compact domain $\mathbf{x}$, there are points ζ^+ and ζ^- in $\mathbf{x}$ where the derivative attains is maximum and minimum, respectively. Thus, for all $x \in \mathbf{x}$, the function value $f(x)$ lies between $t_{\zeta^+}(x)$ and $t_{\zeta^-}(x)$ (see Figure 5.4). Therefore, since $f'(\zeta) \in F'(\mathbf{x})$, the graph of f must intersect the x-axis somewhere within the set

$$Z(\mathbf{x}) \overset{\text{def}}{=} \{t_\zeta(x) = 0: \zeta \in \mathbf{x}\},$$

that is, $x^* \in Z(x)$. But, by construction, $N(x)$ contains $Z(x)$. Hence $x^* \in N(x) \subseteq x$, as claimed.   □

The following MATLAB code implements a rudimentary non-linear solver using the interval Newton operator.

```
01 function x = intervalNewtonSearch(fcnName, x, tol)
02 f = inline(fcnName);
03 x0 = x;                        % Save the original domain.
04 printInterval(x);
05 while (Diam(x) > tol)
06     x = newtonStep(f, x);
07     if  isempty(x)
08         disp('No zeros in the given domain.');
09         return;
10     end
11     printInterval(x);
12 end
13 if (x < x0)                    % Check for strict inclusion.
14     fprintf('There is a unique zero in the final interval.');
15 end
16 end
17
18 function y = newtonStep(f, x)
19 xx = autodiff(x, 'variable'); % Cast to autodiff.
20 fx = f(xx);
21 mx = interval(Mid(x));
22 Nx = mx - f(mx)/getDer(fx);    % Interval Newton operator.
23 y  = x & Nx;                   % Intersection can be empty.
24 end
25
26 function printInterval(x)
27 fprintf('x = [%.17f, %.17f]; rad = %e\n', Inf(x), Sup(x), Diam(x)/2);
28 end
```

Example 5.1.2 *Consider the polynomial*

$$f(x) = -2.001 + 3x - x^3,$$

which has derivative $f'(x) = 3(1 - x^2)$. *If we choose* $x_0 = [-3, -3/2]$, *then* $F'(x_0) = [-24, -15/4]$, *so* $N(x_0)$ *is well-defined, and Theorem 5.4 holds.*

The following output was generated by the short MATLAB program above. Of course, all interval operations are performed with directed rounding.

```
>> intervalNewtonSearch('-2.001 + 3*x - x*x*x', interval(-3.0,-3/2), 1e-15);
x = [-3.00000000000000000, -1.50000000000000000]; rad = 7.500000e-01
x = [-2.14001562500000020, -1.54609999999999981]; rad = 2.969578e-01
x = [-2.14001562500000020, -1.96127739828410874]; rad = 8.936911e-02
x = [-2.00684923964035011, -1.99557058024720901]; rad = 5.639330e-03
x = [-2.00012010448626887, -2.00010360853027747]; rad = 8.247978e-06
x = [-2.00011110289039129, -2.00011110287381610]; rad = 8.287593e-12
x = [-2.00011110288172578, -2.00011110288172445]; rad = 6.661338e-16
x = [-2.00011110288172578, -2.00011110288172489]; rad = 4.440892e-16
There is a unique zero in the final interval.
```

It follows from Theorem 5.4 that the final interval contains the unique zero of f *in the domain* $[-3, -3/2]$. *This result would require some theoretical work to prove in a traditional manner.*

To see how our program handles the case when there are no zeros, we pick a different initial region.

Example 5.1.3 *Using the same function as in the previous example, we take* $x_0 =$ *[3/2, 5/2]. Since* $F'(x_0) = [-57/4, -15/4]$ *does not contain zero, Theorem 5.4 still holds. We now receive the following*[2] *output:*

```
>> intervalNewtonSearch('-2.001+3*x-x*x*x',interval(3/2,5/2),1e-15);
x = [1.50000000000000000, 2.50000000000000000]; rad = 5.000000e-01
x = [1.50000000000000000, 1.74596825396825417]; rad = 1.229841e-01
Warning: The intervals do not intersect.
> In interval.and at 5
  In intervalNewtonSearch>newtonStep at 23
  In intervalNewtonSearch at 6
There are no zeros in the given domain.
```

Thus, according to Theorem 5.5, there are no zeros in the region [3/2, 5/2].

If, however, we would attempt to find a zero by the real-valued Newton's method (as implemented in Section 5.1.2 but with more informative output) with $x_0 = 2 \in$ [3/2, 5/2], the resulting sequence would be

```
>> newtonSearch('-2.001 + 3*x - x*x*x', 2, 1e-15);
 x(1)   =  1.55544444444444441
 x(2)   =  1.29760903104348646
 x(3)   =  1.15474191305680840
 x(4)   =  1.07822336434922272
 x(5)   =  1.03755194336880296
...
 x(200) =  0.98888327717523195
 x(201) =  1.00952822174061674
 x(202) =  0.98736268248266268
 x(203) =  1.00696705047187973
 x(204) =  0.97964847315760384
 x(205) =  0.99813269057521359
...
 x(319) = -2.07869190909348056
 x(320) = -2.00392748732078640
 x(321) = -2.00012077493454887
 x(322) = -2.00011110294408434
 x(323) = -2.00011110288172489
```

This sequence takes ages to settle down, and the first few hundred iterates display an almost erratic behavior near the point $x = 1$. As we already have pointed out, this is no anomaly but an inherent property of the real-valued Newton method.

Exercise 5.6. *Write your own interval Newton routine. To handle more general situations, use a bisection scheme with monotonicity checks to single out subdomains where* $0 \notin F'(x)$.

Exercise 5.7. *An alternative stopping condition is to check for when two consecutive iterates are equal. Due to the finite precision of the computations, this will inform us when the rounding errors prevent further convergence.*

---

[2] As pointed out in Section 2.4.1, the output could be made more elegant by removing the warning statement in the implementation of the base interval class.

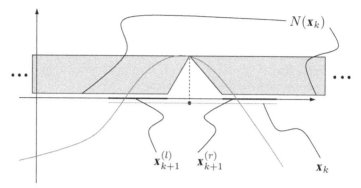

Figure 5.5 One iterate of the extended interval Newton method.

5.1.4 The Extended Interval Newton Method

Using the extended interval division, as described in Section 2.3.4, we can compute

$$N(\mathbf{x}) = \text{mid}(\mathbf{x}) - \frac{f(\text{mid}(\mathbf{x}))}{F'(\mathbf{x})},$$

even when $0 \in F'(\mathbf{x})$. This situation occurs, for example, when the domain $\mathbf{x}$ contains several zeros of f. In a situation where all zeros of a function are sought, it is extremely convenient to be able to handle this type of "division by zero" in a seamless manner.

At any stage of the iteration, the extended interval Newton operator may produce one or two unbounded intervals, depending on the exact position of the interval $F'(\mathbf{x}_k)$. After intersecting with the current interval $\mathbf{x}_k$, the result is therefore one or two compact intervals, that is, we have either

$$N(\mathbf{x}_k) \cap \mathbf{x}_k = \mathbf{x}_{k+1},$$

as usual, or

$$N(\mathbf{x}_k) \cap \mathbf{x}_k = \mathbf{x}_{k+1}^{(l)} \cup \mathbf{x}_{k+1}^{(r)},$$

as illustrated in Figure 5.5. In some (very rare) situations we may have $N(\mathbf{x}_k) = \mathbf{x}_k$, in which case we simply bisect $\mathbf{x}_k$ before continuing. The great advantage with this extension is that it automatically bisects the search domain in exactly the right places. This separates the zeros and allows for local convergence of the interval Newton method. Of course, one can achieve something similar by coupling the interval Newton method with a bisection routine, but then the splitting will not be as effective. The extended method allows us to find *all* zeros of a function within a given domain $\mathbf{x}$. We illustrate this in the following example.

Example 5.1.4 *Consider the function $f(x) = \sin x(x - \cos x)$ used in Example 5.1.1 Again, we will consider the domain $\mathbf{x} = [-10, 10]$. Recall that the function has eight zeros in $\mathbf{x}$: $\{\pm 3\pi, \pm 2\pi, \pm \pi, 0, x^*\}$, where x^* is the unique (positive) zero of $x - \cos x = 0$. Running an extended interval Newton search with tolerance 2^{-10} produces the nine intervals listed below. Note, however, that intervals 5 and*

9 are adjacent. As remarked earlier, this always happens when a zero is located exactly at a bisection point of the domain.

```
Domain                 : [-10, +10]
Tolerance              : 9.765625e-04   (2^-10)
Function calls         : 39
  Unique zero in [-6.28395516124695064,-6.28268172307243855]
  Unique zero in [-9.42494303272712486,-9.42466416085800240]
  Unique zero in [+9.42469385546710114,+9.42492302076889210$]
  Unique zero in [+6.28318404553732801,+6.28318680558727749]
  Maybe a zero in [-0.00022342911344772,+0.00000000000000000]
  Unique zero in [-3.14160246753701244,-3.14158303523038152$]
  Unique zero in [+3.14159233975917740,+3.14159301982135242]
  Unique zero in [$+0.73905272198116201,+0.73913804091868240]
  Maybe a zero in [+0.00000000000000000,+0.00009861499990579]
```

If we change the domain ever so slightly to $x = [-10, 10.001]$, *we get a confirmation that the function indeed has exactly eight zeros in* x:

```
Domain                 : [-10, +10.001]
Tolerance              : 9.765625e-04   (2^-10)
Function calls         : 39
  Unique zero in [-6.28396275881754285,-6.28267595726734562]
  Unique zero in [+9.42469464339922958,+9.42492109675931999]
  Unique zero in [+6.28318411250503672,+6.28318673047858933]
  Unique zero in [-9.42494326359499546,-9.42466402939086834]
  Unique zero in [+3.14159234580190461,+3.14159301300500626]
  Unique zero in [+0.73905039495704095,+0.73914183386738308]
  Unique zero in [-0.00021800677132035,+0.00008258245398134]
  Unique zero in [-3.14160239467736169,-3.14158310518228356]
```

Exercise 5.8. *Write your own extended interval Newton routine. This requires that you implement the extended interval division as described in Section 2.3.4.*

5.1.5 The Krawczyk Method

An attractive alternative to the interval Newton method is that of Krawczyk (see [Kr69]), which avoids the perils of having to divide by the enclosure $F'(x)$ when computing the Newton iterates. In light of what we have done so far, the idea is very simple: Assuming that $f \in C^1(x, \mathbb{R})$ has a zero $x^* \in x$, we Taylor expand around x^* to get $f(x) = f(x^*) + f'(\zeta)(x - x^*) = f'(\zeta)(x - x^*)$ for some ζ between x and x^*. Next, we multiply the expansion by some (finite) constant C,

$$Cf(x) = Cf'(\zeta)(x - x^*),$$

and add $x^* - x$ to both sides of the equation:

$$x^* - x + Cf(x) = x^* - x + Cf'(\zeta)(x - x^*).$$

After a rearrangement, this reduces to

$$x^* = x - Cf(x) - \left(1 - Cf'(\zeta)\right)(x - x^*).$$

Now, although we know neither the zero x^* nor the point ζ, we do know that both points belong to the domain x. Thus we can enclose the zero via

$$x^* \in x - Cf(x) - \left(1 - CF'(x)\right)(x - x) \stackrel{\text{def}}{=} K(x, x, C).$$

We have now proved that any zero $x^* \in x$ of f is also enclosed by the Krawczyk operator $K(x, x, C)$ for any $x \in x$ and C finite. At this point, the benefit of Krawczyk's formulation should be apparent: we no longer have to divide by $F'(x)$, which (as we have seen) is a potential danger. Good choices for x and C are $x = m = \text{mid}(x)$ and $C = 1/f'(m)$, which results in the operator

$$K(x) \stackrel{\text{def}}{=} K(x, m, 1/f'(m)) = m - \frac{f(m)}{f'(m)} - \left(1 - \frac{F'(x)}{f'(m)}\right)[-r, r],$$

where we use the notation $r = \text{rad}(x)$.

Given an initial search region x_0 for a zero x^*, we define the sequence of intervals

$$x_{k+1} = K(x_k) \cap x_k, \quad k = 0, 1, 2, \ldots \tag{5.2}$$

forming the *Krawczyk iterates* of x_0. As long as $f'(m_k) \neq 0$ for all $k \in \mathbb{N}$, the sequence is well-defined. Analogous to Theorems 5.4 and 5.5, we have the following theorem.

THEOREM 5.9. *Assume that $K(x)$ is well-defined. Then the following statements hold:*

(1) if x contains a zero x^* of f, then so does $K(x) \cap x$;
(2) if $K(x) \cap x = \emptyset$, then x contains no zeros of f;
(3) if $K(x) \subseteq \text{int}(x)$, then x contains exactly one zero of f.

Proof. Part (1) holds true by construction. Part (2) follows immediately from the contrapositive statement of (1). To prove part (3), we note that $K(x) \subseteq \text{int}(x)$ implies that the width of the Krawczyk image satisfies $w(K(x)) < w(x) = w([-r, r])$. On the other hand, we also have

$$w(K(x)) = w\left(\left(1 - \frac{F'(x)}{f'(m)}\right)[-r, r]\right),$$

which implies that $F'(x)/f'(m) \subset (0, 2)$. Since $K(x)$ is well-defined, we know that $f'(m) \neq 0$. It follows that $0 \notin F'(x)$, which means that f' is non-zero on x, so f can have at most one zero in x. To see that f indeed has a zero in x, we note that the Krawczyk operator is simply the midpoint form of the operator $G(x, x) = x - f(x)/f'(\text{mid}(x))$. Following Section 3.2, we have

$$G_m(x, x) = m - \frac{f(m)}{f'(m)} + \left(1 - \frac{F(x)}{f'(m)}\right)(x - m) = K(x).$$

Therefore, $G(x, x) \in K(x) \subseteq \text{int}(x)$ for all $x \in x$. By the Intermediate Value Theorem, this implies that G (and thus f) has a zero in x. This completes the proof. □

We end this section with a simple implementation of the Krawczyk method, merged with a bisection scheme (see Listing 5.2). Although this code is not optimal with regard to the number of calls to the Krawczyk operator K, its simple structure has some merit. An obvious improvement would be to use automatic differentiation rather than hand-coding the derivative.

We will use the same function $f(x) = \sin x(x - \cos x)$ as in Examples 5.1.1 and 5.1.4 Note that the extended interval Newton method is more efficient than the Krawczyk method, in that it requires fewer function evaluations. In fact, the Newton methods require one derivative evaluation per function evaluation, whereas for the Krawczyk operator the ratio is $2:1$.

```
Domain            : [-10, 10]
Tolerance         : 9.765625e-04  (2^-10)
Function calls    : 89
  Unique zero in [-9.424778380712580,-9.424777714952242]
  Unique zero in [-6.283186564304483,-6.283184617341837]
 Maybe a zero in [-3.141722588593495,-3.141306012318185]
 Maybe a zero in [-0.000000581733623,+0.000000000000000]
 Maybe a zero in [+0.000000000000000,+0.000253365746756]
  Unique zero in [+0.738912754470971,+0.739439826304897]
  Unique zero in [+3.141585088281115,+3.141614165426731]
  Unique zero in [+6.283180363246271,+6.283193001393917]
  Unique zero in [+9.424777809870829,+9.424778213686277]
```

All eight zeros can be securely enclosed if we decrease the tolerance and shift the domain. This only incurs a small increase in the computational cost.

```
Domain            : [-10, 10.001]
Tolerance         : 0.0001
Function calls    : 93
  Unique zero in [-9.424778381078896,-9.424777714652166]
  Unique zero in [-6.283186567789095,-6.283184616617528]
  Unique zero in [-3.141592664446003,-3.141592643554570]
  Unique zero in [-0.000000642135820,+0.000000633776171]
```

```
Unique zero in [+0.739085090377269,+0.739085180166543]
Unique zero in [+3.141585146865033,+3.141614053175048]
Unique zero in [+6.283180355257999,+6.283192959093360]
Unique zero in [+9.424777813489889,+9.424778211462723]
```

Listing 5.2. A recursive Krawczyk/bisection scheme using the PROFIL/BIAS package [PrBi]

```
 1 #include <iostream>
 2 #include "Interval.h"
 3 #include "Functions.h"
 4 using namespace std;
 5 typedef INTERVAL (*pfcn)(INTERVAL, int);
 6
 7 INTERVAL function(INTERVAL X, int n) {
 8   if ( n == 0 ) return Sin(X)*(X - Cos(X));
 9   else           return Sin(X)*(1 + Sin(X)) + Cos(X)*(X - Cos(X));
10 }
11
12 INTERVAL F  (INTERVAL X) { return function(X,0); }
13 INTERVAL DF (INTERVAL X) { return function(X,1); }
14
15 INTERVAL K (pfcn f, INTERVAL X)
16 {
17   INTERVAL   m(Mid(X));
18   INTERVAL DFm(DF(m));
19   return m - F(m)/DFm + (1 - DF(X)/DFm)*(x - m);
20 }
21
22 void krawczyk(pfcn f, INTERVAL X, double Tol) {
23   INTERVAL KX = K(f, X);
24   if(Intersection(KX, KX, X)) {
25     if (Diam(KX) < Tol) {
26       if (KX < X)
27         cout << " Unique zero in " << KX << endl;
28       else
29         cout << "Maybe a zero in " << KX << endl;
30     }
31     else {
32       krawczyk(f, INTERVAL(Inf(KX), Mid(KX)), Tol);
33       krawczyk(f, INTERVAL(Mid(KX), Sup(KX)), Tol);
34     }
35   }
36 }
37
38 int main(int argc, char * argv[])
39 {
40   INTERVAL   X(atof(argv[1]), atof(argv[2]));
41   double   Tol(atof(argv[3]));
42
43   krawczyk(function, X, Tol);
44   return 0;
45 }
```

5.2 OPTIMIZATION

In this section, we will focus on the task of optimization. In essence, given a function $f: D \to \mathbb{R}$, we wish to find the minimal value y^* of f over the domain D:

$$y^* = y^*(f; D) = \inf\{f(x): x \in D\}, \qquad (5.3)$$

as well as the set of points in D where this minimum is attained:

$$\mathbb{E}^* = \mathbb{E}^*(f; D) = \{x^* \in D: f(x^*) = y^*\}. \qquad (5.4)$$

Note that finding the maximal value of a function is achieved by minimizing $-f$. In what follows, we will assume that f is continuous on its domain D, which is assumed to be compact. This implies that $\mathbb{E}^*$ is non-empty, that is, f attains its minimum on D. Note that the set of extremal points $\mathbb{E}^*$ may very well contain an entire line segment. This happen, for example, when f is constant and $D = x$.

We will employ techniques from interval analysis to enclose both y^* and $\mathbb{E}^*$. Starting with the extremal value y^*, note that (5.3) can be expressed as

$$y^* = \inf\{y: y \in R(f; D)\}. \qquad (5.5)$$

Therefore, if we can enclose $R(f; D)$ tightly, then we immediately have a tight enclosure of y^*. In fact, optimization is less strenuous than finding a range enclosure, since we need only focus our attention on the lower endpoint of the range $R(f; D)$.

The techniques we will employ are very similar to the zero-finding techniques presented earlier. By excluding subsets of D, we will attempt to solve the problem

$$f(x) \le \tilde{y} \qquad (5.6)$$

but with a varying value for $\tilde{y}$, which is the current upper bound for y^*. During the process of solving (5.6) we may come across a smaller value for $\tilde{y}$, in which case its value is decreased, and the excluding process continues. The goal is to make $\tilde{y}$ as close to y^* as possible, since then solving (5.6) generates the desired set

$$\mathbb{E}^* = \{x^* \in D: f(x^*) \le y^*\} = \{x^* \in D: f(x^*) = y^*\}. \qquad (5.7)$$

Realistically, we can hope to find an interval $\mathbf{y}^*$ containing y^*, and an interval enclosure $\cup_i \mathbf{x}_i^*$ of the set $\mathbb{E}^*$. Following the spirit of [HH95],[3] we will use several criteria for discarding subintervals from our search: the midpoint method, the monotonicity test, and the concavity test. We will from now on assume that the domain D is an interval $\mathbf{x}$. More complicated domains can be treated by considering unions of intervals.

5.2.1 The Midpoint Method

To initialize our search, we set $\mathbb{E}^{(0)} = \mathbf{x}^{(0)} = \mathbf{x}$, and $\tilde{y}^{(0)} = +\infty$. It is then clear that $y^* \le \tilde{y}^{(0)}$, and thus that $\mathbb{E}^* \subseteq \{x \in \mathbf{x}: f(x) \le \tilde{y}^{(0)}\}$.

---

[3] See also the books [Ke96] and [WH03], as well as the comprehensive survey article [Ne04].

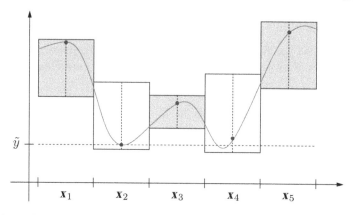

Figure 5.6 Eliminating subintervals x_1, x_3, and x_5 with the midpoint method.

Now, suppose that we have arrived at stage k of our optimization process. At this point, we have an upper bound for the minimal function value $y^* \leq \tilde{y}^{(k)}$ and the enclosure

$$\mathbb{E}^* \subseteq \{x \in \mathbf{x}: f(x) \leq \tilde{y}^{(k)}\} \subseteq \mathbb{E}^{(k)} = \bigcup_{i=1}^{N_k} \mathbf{x}_i^{(k)}$$

of the minimizing set. For each $i = 1, \ldots, N_k$, we do the following: First, we compute the enclosure $\mathbf{y}_i = F(\mathbf{x}_i^{(k)})$ and check if $\tilde{y}^{(k)} < \underline{y}_i$. If this is the case, then the set $\mathbf{x}_i^{(k)}$ is eliminated from the search (see Figure 5.6). Otherwise, we continue by computing the midpoint sample $m_i = f(\text{mid}(\mathbf{x}_i^{(k)}))$, which is compared to $\tilde{y}^{(k)}$. If $m_i < \tilde{y}^{(k)}$, we set $\tilde{y}^{(k)} = m_i$. Also, we bisect the subdomain $\mathbf{x}_i^{(k)}$ and add the two halves to $\mathbb{E}^{(k+1)}$. When we have looped through all i, we have arrived at stage $k+1$ and set $\tilde{y}^{(k+1)} = \tilde{y}^{(k)}$. The entire process is repeated until some stopping criterion has been met. This could be, for example, when the subdomains are sufficiently small, that is, when

$$\max \left\{\text{rad}(\tilde{\mathbf{x}}): \tilde{\mathbf{x}} \in \mathbb{E}^{(k)}\right\} \leq \text{TOL}, \tag{5.8}$$

or when the local range enclosures are sufficiently tight, that is, when

$$\max \left\{\text{rad}(F(\tilde{\mathbf{x}})): \tilde{\mathbf{x}} \in \mathbb{E}^{(k)}\right\} \leq \text{TOL}. \tag{5.9}$$

Using the stopping criteria (5.8) will produce an upper bound for the global minimum: $y^* \leq \tilde{y} = \tilde{y}^{(k)}$, as well as an enclosure $\mathbb{E}^{(k)}$ of the set $\{x \in \mathbf{x}: f(x) \leq \tilde{y}\}$, which of course encloses $\mathbb{E}^*$. We are, however, not provided with any a priori information regarding the quality of the computed upper bound (i.e., a bound on $y^* - \tilde{y}$). On the other hand, we know that the algorithm will terminate within an explicit number of loops. This choice is suitable when we only need to know that the global minimum is smaller than some specific value or that it is attained within some specific set.

Using the stopping criteria (5.9) has the advantage of producing an enclosure of global minimum: $y^* \in \mathbf{y}^* = [\tilde{y} - \text{TOL}, \tilde{y}]$. A disadvantage is that we no longer

Listing 5.3.   A midpoint optimization procedure using the PROFIL/BIAS package [PrBi]

```
 1  void valGlobMin(INTERVAL domain, double TOL)
 2  {
 3    double globalMin = Machine::PosInfinity;
 4    List<INTERVAL> minimizingList, domainList;
 5    domainList += domain;
 6    while( !IsEmpty(domainList) ) {
 7      INTERVAL localX = Pop(domainList);
 8      INTERVAL localY = F(localX);
 9      if ( globalMin >= Inf(localY) ) {
10        double midY = Sup(F(Hull(Mid(localX))));
11        if ( globalMin > midY )
12          globalMin = midY;
13        if ( Rad(localY) > TOL )          // Or 'Rad(localX) > TOL'.
14          splitAndStore(localX, domainList);
15        else
16          minimizingList += localX;
17      }
18    }
19    INTERVAL Y = boundMinimum(minimizingList, globalMin);
20    cout << "Global minimum in: " << Mid(Y) << " +- "
21         << Rad(Y) << endl;
22    cout << "Attained within  : " << minimizingList  << endl;
23  }
```

know that the algorithm will terminate within an explicit number of loops. This choice is suitable when we really need to know the value and location of the global minimum.

We stress that any point $\check{x}_i^{(k)}$ of the subinterval $\mathbf{x}_i^{(k)}$ could be used for taking the sample $m_i = f(\check{x}_i^{(k)})$. Consequently, we are not worried about the fact that the midpoint of an interval may not be exactly representable: any point within the subinterval will suffice for our needs. In some situations it may be beneficial to sample f at several points of $\mathbf{x}_i^{(k)}$ and then use their collective minimum as the value of m_i. The procedure just described can be programmed (in C++) as in Listing 5.3.

Note that on line 10, we compute the largest element of the set $F([\check{x}, \check{x}])$, where $\check{x}$ is the floating point evaluation of the midpoint of localX. The reason why we want the *largest* element is that we do not want to underestimate the global minimum.

Identifying $\tilde{y}^{(k)}$ with globalMin and $\mathbb{E}^{(k)}$ with minimizingList, the function call on line 19 computes the interval $\mathbf{y} = \mathbf{z} \cap [-\infty, \tilde{y}^{(k)}]$, where

$$\underline{z} = \min\{\inf\{F(\tilde{\mathbf{x}})\} \colon \tilde{\mathbf{x}} \in \mathbb{E}^{(k)}\} \quad \text{and} \quad \overline{z} = \min\{\sup\{F(\tilde{\mathbf{x}})\} \colon \tilde{\mathbf{x}} \in \mathbb{E}^{(k)}\}.$$

Thus the resulting enclosure $\mathbf{y}$ may be tighter than could be expected by the specified tolerance.

Example 5.2.1 *Running the coded algorithm with $f(x) = \cos x$ and TOL$= 2^{-10}$ over the domain $\mathbf{x} = [-15, 15]$ produces the following output:*

```
Domain              : [-15,15]
Global minimum in: -0.999988347965415 +- 1.16520345848636e-05
Attained within  :
   1: [-9.433593750000000e+00,-9.375000000000000e+00]
   2: [-3.164062500000000e+00,-3.105468750000000e+00]
   3: [+3.105468750000000e+00,+3.164062500000000e+00]
   4: [+9.375000000000000e+00,+9.433593750000000e+00]
Tolerance           : 9.765625000000000e-04  (2^-10)
Function calls      : 122
```

Decreasing the tolerance to 2^{-40} produces several more intervals. Note, however, that now the subintervals labeled 2 and 3 are adjacent, as are those labeled 4 and 5. This has to do with the symmetry of the function with respect to the bisection points. It is easy to write a small routine that merges adjacent intervals before sending the list to the output.

```
Domain              : [-15,15]
Global minimum in: -0.999999999999998 +- 2.22044604925031e-15
Attained within  :
   1: [-9.424778223037720e+00,-9.424776434898376e+00]
   2: [-3.141594529151917e+00,-3.141592741012573e+00]
   3: [-3.141592741012573e+00,-3.141590952873230e+00]
   4: [+3.141590952873230e+00,+3.141592741012573e+00]
   5: [+3.141592741012573e+00,+3.141594529151917e+00]
   6: [+9.424776434898376e+00,+9.424778223037720e+00]
Tolerance           : 9.094947017729282e-13   (2^-40)
Function calls      : 343
```

It should be pointed out that decreasing the tolerance is meaningless once the maximum accuracy of the floating point computation of f is reached. Repeating Example 5.2.1 with tolerance 2^{-50} produces 16 intervals, and with tolerance 2^{-51}, the program never terminates.

Exercise 5.10. *Write a program implementing the midpoint method, but use the stopping criteria (5.8). How does decreasing the tolerance now affect the output?*

Exercise 5.11. *Write and add the merging routine described in Example 5.2.1 to your program.*

5.2.2 The Monotonicity Test

Checking if f is monotone on subintervals is also an effective way of detecting areas where the global minimum cannot be obtained. This requires that we have an interval extension of f', which we will call F'. If we, at some stage of our search,

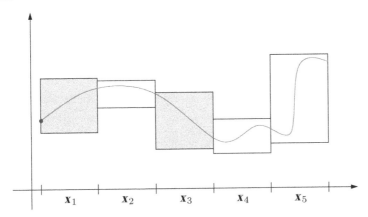

Figure 5.7 Eliminating subinterval x_3, and shrinking x_1 to its left endpoint with the monotonicity test.

have a subinterval $\mathbf{x}_i^{(k)} \in \mathbb{E}^{(k)}$ such that $0 \in F'(\mathbf{x}_i^{(k)})$, then we bisect $\mathbf{x}_i^{(k)}$ and add the two halves to $\mathbb{E}^{(k+1)}$. If, on the other hand, $0 \notin F'(\mathbf{x}_i^{(k)})$, there are two situations that need separate treatment. In the first case, $\mathbf{x}_i^{(k)}$ belongs to the interior of $\mathbf{x}^{(0)}$, in which case we can delete the entire set $\mathbf{x}_i^{(k)}$ from our search. In the second case, $\mathbf{x}_i^{(k)}$ has at least one endpoint in common with $\mathbf{x}^{(0)}$. If f is increasing on $\mathbf{x}_i^{(k)}$, and if $\underline{x}_i^{(k)} = \underline{x}^{(0)}$, then we shrink $\mathbf{x}_i^{(k)}$ to the point $\underline{x}_i^{(k)}$, which is added to $\mathbb{E}^{(k+1)}$. If f is decreasing on $\mathbf{x}_i^{(k)}$, and if $\overline{x}_i^{(k)} = \overline{x}^{(0)}$, then we shrink $\mathbf{x}_i^{(k)}$ to the point $\overline{x}_i^{(k)}$, which is added to $\mathbb{E}^{(k+1)}$. A typical application of the monotonicity test is illustrated in Figure 5.7.

Listing 5.4. The monotonicity test

```
1  action monotoneTest(INTERVAL &newX, INTERVAL oldX,
2  INTERVAL domain)
3  {
4    INTERVAL df = DF(X);
5    if ( Mig(df) == 0 ) {
6      newX = oldX;
7      return PRESERVE;
8    }
9    else if ((Inf(oldX) == Inf(domain)) && (0.0 < Inf(df)))
10     newX = Hull(Inf(oldX));
11   else if ((Sup(oldX) == Sup(domain)) && (Sup(df) < 0.0))
12     newX = Hull(Sup(oldX));
13   else
14      return DELETE;
15   return SHRINK;
16 }
```

Implementing this test is straightforward. In Listing 5.4 a simple C++ routine is presented that preserves, shrinks, or deletes a subinterval (denoted `oldX` in the code), depending on the monotonicity properties of f. This small routine is merged with `valGlobMin` as illustrated in Listing 5.5.

Listing 5.5.   The monotonicity test inserted in the code proper

```
 1  ...
 2    while( !IsEmpty(domainList) ) {
 3      INTERVAL oldX = First(domainList);
 4      RemoveCurrent(domainList);
 5      INTERVAL localX;
 6      if (DELETE == monotoneTest(localX, oldX, domain))
 7        continue;
 8      INTERVAL localY = F(localX);
 9      if ( globalMin >= Inf(localY) ) {
10  ...
```

When the test deletes a subinterval, the instruction continue returns the program flow to the top of the while-loop, that is, a new element of domainList is examined.

Example 5.2.2 *Continuing Example 5.2.1, we repeat the runs but this time using the additional monotonicity test.*

Running the updated algorithm with $f(x) = \cos x$ *and* TOL $= 2^{-10}$ *over the domain* $\mathbf{x} = [-15, 15]$ *produces the following output:*

```
Domain             : [-15,15]
Global minimum in: -0.999988347965415 +- 1.16520345848636e-05
Attained within  :
  1: [-9.433593750000000e+00,-9.375000000000000e+00]
  2: [-3.164062500000000e+00,-3.105468750000000e+00]
  3: [+3.105468750000000e+00,+3.164062500000000e+00]
  4: [+9.375000000000000e+00,+9.433593750000000e+00]
Tolerance          : 9.765625000000000e-04   (2^{-10})
Function calls     : 76
Derivative calls   : 63
```

Without the monotonicity test, 122 interval function calls were required. This should be compared to the $76 + 63 = 139$ *calls now required. With the monotonicity test, however, decreasing the tolerance to* 2^{-40} *does not produce additional intervals.*

```
Domain             : [-15,15]
Global minimum in: -0.999999999999998 +- 2.22044604925031e-15
Attained within  :
  1: [-9.424778223037720e+00,-9.424776434898376e+00]
  2: [-3.141592741012573e+00,-3.141590952873230e+00]
  3: [+3.141590952873230e+00,+3.141592741012573e+00]
  4: [+9.424776434898376e+00,+9.424778223037720e+00]
```

```
Tolerance         : 9.094947017729282e-13   (2^{-40})
Function calls    : 196
Derivative calls  : 183
```

Again, the number of calls $196 + 183 = 379$ is slightly larger than the number required (343) without the monotonicity test. An advantage is that we now obtain a tighter enclosure of the minimizing set $\mathbb{E}^*$.

5.2.3 The Convexity Test

Digging even deeper for helpful information, the second derivative f'' also provides a powerful means of detecting areas where the global minimum cannot be obtained. This, of course, requires that we have an interval extension of f'', which we will call F''. If the global minimum is attained within the interior of $\mathbf{x}^{(0)}$, that is, if $\mathbb{E}^* \overset{\circ}{\subset} \mathbf{x}^{(0)}$, then for all $x^* \in \mathbb{E}^*$, we have $f''(x^*) \geq 0$. Therefore, if we, at some stage of our search, have a subinterval $\mathbf{x}_i^{(k)} \in \mathbb{E}^{(k)}$ such that $\sup\{F''(\mathbf{x}_i^{(k)})\} < 0$, then $\mathbf{x}_i^{(k)}$ can be deleted from the search, unless $\mathbf{x}_i^{(k)}$ has a boundary point in common with $\mathbf{x}^{(0)}$. If the latter happens, we shrink $\mathbf{x}_i^{(k)}$ to the set $\mathbf{x}_i^{(k)} \cap \{\underline{x}^{(0)}, \overline{x}^{(0)}\}$, which is added to $\mathbb{E}^{(k+1)}$. A typical application of the convexity test is illustrated in Figure 5.8.

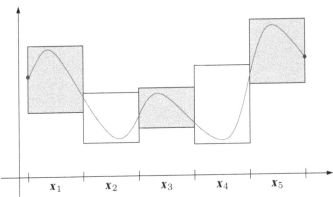

Figure 5.8 Shrinking the subintervals $\mathbf{x}_1$ and $\mathbf{x}_5$, and eliminating $\mathbf{x}_3$ with the convexity test.

Much like the monotonicity test, it is very easy to implement the necessary checks for convexity. The core C++ routine preserves, shrinks, or deletes a subinterval, depending on the convexity properties of f.

Exercise 5.12. *Modify your code so that it also implements the convexity test. On termination, the program should print the percentages of the initial interval deleted by the various tests. Note that this will depend on which order they are performed.*

Exercise 5.13. *Assuming that f is twice continuously differentiable, it is possible to employ the interval Newton methods for f' on subintervals where f'' does not vanish. This will determine whether f has a stationary point in the subinterval. Modify your code so that it implements this additional test.*

Listing 5.6. The convexity test. The returned value informs about the action taken on the domain under scrutiny

```
1  action convexTest(INTERVAL &newX, INTERVAL oldX, INTERVAL domain)
2  {
3    INTERVAL ddf = DDF(oldX);
4    if ( Sup(ddf) < 0.0 ) {
5      if ( Inf(oldX) == Inf(domain) )
6        newX = Hull(Inf(oldX));
7      else if ( Sup(oldX) == Sup(domain) )
8        newX = Hull(Sup(oldX));
9      else
10       return DELETE;
11     return SHRINK;
12   }
13   return PRESERVE;
14 }
```

5.3 QUADRATURE

We now turn our attention to the problem of computing, or enclosing, definite integrals, that is, integrals of the type

$$I = \int_a^b f(x)dx. \tag{5.10}$$

Today there are myriad numerical methods designed to produce efficient and accurate estimates of integrals of type (5.10). We will briefly mention a few of the most elementary such methods: the *midpoint* method, the *trapezoid* method, and *Simpson's* method.

All integration methods rely on the basic fact that the integral operator is additive with respect to the domain, that is, for any partition $a = x_0 \le x_1 \le \cdots \le x_N = b$, we have

$$\int_a^b f(x)dx = \sum_{i=1}^N \int_{x_{i-1}}^{x_i} f(x)dx.$$

By taking a sufficiently fine partition, it is likely that the integrand f can be well approximated by low-order polynomials p_i over each subinterval $[x_{i-1}, x_i]$, which results in the approximation

$$I = \int_a^b f(x)dx = \sum_{i=1}^N \int_{x_{i-1}}^{x_i} f(x)dx$$

$$\approx \sum_{i=1}^N \int_{x_{i-1}}^{x_i} p_i(x)dx = \sum_{i=1}^N \left(P_i(x_i) - P_i(x_{i-1}) \right).$$

$$\tag{5.11}$$

Here, the polynomials P_i satisfy the relation $P_i'(x) = p_i(x)$.

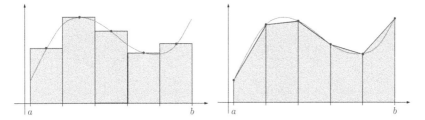

Figure 5.9 (left) The midpoint method; (right) the trapezoidal method.

For the midpoint method, we take $p_i(x) = f(\frac{x_{i-1}+x_i}{2})$, that is, on each subinterval the corresponding p_i is a constant equal to the function value at the midpoint of $[x_{i-1}, x_i]$; the latter we will denote by $\tilde{x}_i$. The trapezoid method is obtained by taking $p_i(x) = f(x_{i-1})(1-s_i(x))+f(x_i)s_i(x)$ with $s_i(x) = (x-x_{i-1})/(x_i-x_{i-1})$, that is, on each subinterval the corresponding p_i is the linear function passing from $f(x_{i-1})$ to $f(x_i)$. Both methods are illustrated in Figure 5.9.

Simpson's method approximates f, restricted to $[x_{i-1}, x_i]$, with a second degree polynomial that coincides with f at the points[4] x_{i-1}, $\tilde{x}_i$, and x_i.

Assuming that the partition is uniform, that is, that $x_i = a + ih$, where $h = (b-a)/N$ is the step-size, there are elegant formulas for the methods just described.

$$I \approx M_a^b(f, N) \overset{\text{def}}{=} h \sum_{i=1}^{N} f(\tilde{x}_i)$$

$$I \approx T_a^b(f, N) \overset{\text{def}}{=} h \left(\frac{f(x_0)}{2} + \sum_{i=1}^{N-1} f(x_i) + \frac{f(x_N)}{2} \right) \tag{5.12}$$

$$I \approx S_a^b(f, N) \overset{\text{def}}{=} \frac{1}{3} T_a^b(f, N) + \frac{2}{3} M_a^b(f, N).$$

The code in Listing 5.7 illustrates how easily one can implement the three methods.

It is not hard to obtain explicit formulas for the integral errors, assuming that the integrand is sufficiently smooth ($f \in C^2([a, b])$ for the midpoint and trapezoidal methods, and $f \in C^4([a, b])$ for Simpson's method). Rather surprisingly, it turns out that the midpoint method is twice as accurate as the trapezoidal rule (for sufficiently small h). More precisely, for some $\zeta_M, \zeta_T, \zeta_S \in [a, b]$, we have

$$\int_a^b f(x)dx - M_a^b(f, N) = \frac{b-a}{24}h^2 f''(\zeta_M) \overset{\text{def}}{=} \Delta M_a^b(f, N),$$

$$\int_a^b f(x)dx - T_a^b(f, N) = -\frac{b-a}{12}h^2 f''(\zeta_T) \overset{\text{def}}{=} \Delta T_a^b(f, N), \tag{5.13}$$

$$\int_a^b f(x)dx - S_a^b(f, N) = -\frac{b-a}{2880}h^4 f^{(4)}(\zeta_S) \overset{\text{def}}{=} \Delta S_a^b(f, N).$$

---

[4]Our definition is not entirely standard. It is more common to sample at three consecutive grid points x_{i-1}, x_i, and x_{i+1}. This, however, requires that N is even.

Table 5.1 Integral estimates of $\int_{-2}^{2} \sin(\cos(e^x))dx$.

N	$M_{-2}^2(f,N)$	$T_{-2}^2(f,N)$	$S_{-2}^2(f,N)$	$\Delta M_a^b(f,N)$	$\Delta T_a^b(f,N)$	$\Delta S_a^b(f,N)$
1	2.0575810	2.5399605	2.2183742	$-7.189 \cdot 10^{-1}$	$-1.201 \cdot 10^{0}$	$-8.797 \cdot 10^{-1}$
2	0.0257958	2.2987708	0.7834541	$+1.313 \cdot 10^{0}$	$-9.601 \cdot 10^{-1}$	$+5.552 \cdot 10^{-1}$
4	1.2556376	1.1622833	1.2245195	$+8.303 \cdot 10^{-2}$	$+1.764 \cdot 10^{-1}$	$+1.141 \cdot 10^{-1}$
8	1.4089676	1.2089605	1.3422986	$-7.030 \cdot 10^{-2}$	$+1.297 \cdot 10^{-1}$	$-3.630 \cdot 10^{-3}$
16	1.3527757	1.3089641	1.3381718	$-1.411 \cdot 10^{-2}$	$+2.970 \cdot 10^{-2}$	$+4.969 \cdot 10^{-4}$
32	1.3425891	1.3308699	1.3386827	$-3.920 \cdot 10^{-3}$	$+7.799 \cdot 10^{-3}$	$-1.398 \cdot 10^{-5}$
64	1.3396402	1.3367295	1.3386699	$-9.715 \cdot 10^{-4}$	$+1.939 \cdot 10^{-3}$	$-1.235 \cdot 10^{-6}$
128	1.3389108	1.3381848	1.3386688	$-2.421 \cdot 10^{-4}$	$+4.839 \cdot 10^{-4}$	$-8.039 \cdot 10^{-8}$
256	1.3387292	1.3385478	1.3386687	$-6.046 \cdot 10^{-5}$	$+1.209 \cdot 10^{-4}$	$-5.068 \cdot 10^{-9}$
512	1.3386838	1.3386385	1.3386687	$-1.511 \cdot 10^{-5}$	$+3.022 \cdot 10^{-5}$	$-3.174 \cdot 10^{-10}$
1024	1.3386725	1.3386612	1.3386687	$-3.778 \cdot 10^{-6}$	$+7.556 \cdot 10^{-6}$	$-1.988 \cdot 10^{-11}$

Listing 5.7.   A straightforward C++ implementation of the integration schemes

```cpp
1  #include <iostream>
2  using namespace std;
3  typedef double (*pfcn)(double);
4
5  double midpoint(pfcn f, double a, double b, int N) {
6     double h   = (b - a)/N;
7     double sum = f(a + 0.5*h);
8     for (int i = 1; i < N; i++)
9        sum += f(a + (i + 0.5) * h);
10    return h*sum;
11 }
12
13 double trapezoid(pfcn f, double a, double b, int N) {
14    double h   = (b - a)/N;
15    double sum = (f(a) + f(b))*0.5;
16    for (int i = 1; i < N; i++)
17       sum += f(a + i*h);
18    return h*sum;
19 }
20
21 double simpsons(pfcn f, double a, double b, int N) {
22 return (2*midpoint(f, a, b, N) + trapezoid(f, a, b, N))/3;
23 }
24
25 double integrand(double x) { return sin(cos(exp(x))); }
26
27 int main(int argc, char * argv[])
28 {
29    double a(atof(argv[1])), b(atof(argv[2]));
30    int    N(atoi(argv[3]));
31
32    cout <<  midpoint(integrand, a, b, N) << endl;
33    cout << trapezoid(integrand, a, b, N) << endl;
34    cout <<  simpsons(integrand, a, b, N) << endl;
35
36    return 0;
37 }
```

Of course, the points ζ_M, ζ_T, and ζ_S are unknown. Nevertheless, if we can somehow bound the derivatives appearing in (5.13) over the domain $[a, b]$, we obtain upper bounds on the errors. In the next section, this type of approach will be studied in greater detail.

Example 5.3.1 *Using the methods described in this section, let us approximate the integral $I = \int_{-2}^{2} f(x)dx$, where $f(x) = \sin(\cos(e^x))$. For convenience, we will take $N = 2^i$ for $i = 0, \ldots, 10$. The results are presented in Table 5.1. From an auxilliary computation, we happen to know that $I \in 1.33866870740_{18}^{20}$. This allows us to compute the integration errors, which are also displayed in the table.*

Our theory is clearly confirmed by the numerics: Asymptotically, each time we double N, we see both the midpoint and trapezoidal errors decrease by a factor 4.

Also, the Simpson error decreases by a factor 16, as predicted by (5.13). We also note that the trapezoidal error is roughly twice as large as the midpoint error, in good agreement with (5.13).

5.3.1 Enclosure Methods

Interval analysis provides an elegant means for approximating definite integrals. If f admits a well-defined interval extension F over the integration domain $[a, b]$, then, by Theorem 3.8, we have the naive enclosure

$$I \in w([a, b])F([a, b]),$$

where we let $w([a, b]) = b - a$ denote the *width* of an interval. Of course, in most cases this will produce a terrible estimate of I: the resulting interval will most likely be very wide. As mentioned earlier, the integral operator is additive with respect to the domain. Thus, given any partition $a = x_0 \leq x_1 \leq \cdots \leq x_N = b$, we can decompose the integration domain into the N non-overlapping subintervals

$$[a, b] = [x_0, x_1] \cup [x_1, x_2] \cup \cdots \cup [x_{N-1}, x_N]$$

and bound the range of f over each subinterval separately. This produces the enclosure

$$I = \sum_{i=1}^{N} \int_{x_{i-1}}^{x_i} f(x)dx \in \sum_{i=1}^{N} w([x_{i-1}, x_i])F([x_{i-1}, x_i]). \qquad (5.14)$$

If the integrand f is Lipschitz on $[a, b]$, then we can use Theorem 3.10 and deduce that the width of the enclosure (5.14) is at most proportional to $\max_i w([x_{i-1}, x_i])$.

The most straightforward approach is to split the domain of integration into N equally wide subintervals: we set $h = (b - a)/N$ and $x_i = a + ih$, $i = 0, \ldots, N$. This produces the enclosure

$$I_a^b(f, N) \stackrel{\text{def}}{=} h \sum_{i=1}^{N} F([x_{i-1}, x_i]), \qquad (5.15)$$

which satisfies $w(I_a^b(f, N)) = \mathcal{O}(1/N)$. An implementation of this approach is illustrated in Listing 5.8. Note that on line 14 we use an interval-valued diameter in order to get a valid enclosure of the local Riemann term.

Example 5.3.2 *Approximate the integral $I = \int_{-2}^{2} f(x)dx$, where $f(x) = \sin(\cos(e^x))$ by computing the enclosures $I_{-2}^{2}(f, N)$ for $N = 1, 100, 10,000$, and $1,000,000$.*

The results are presented in Table 5.2. Note that, starting from $N = 100$, we clearly see the reciprocal correspondence between the number of subintervals and the enclosure widths.

From this example, it should be clear that method (5.15) is *absolutely not* the right way to go about computing integral enclosures. The enclosure method (5.15)

Table 5.2 Integral enclosures of $\int_{-2}^{2} \sin(\cos(e^x))dx$ using (5.15)

N	$I_{-2}^{2}(f, N)$	$w(I_{-2}^{2}(f, N))$
10^0	$[-3.36588, 3.36588]$	$6.73177 \cdot 10^0$
10^2	$[1.26250, 1.41323]$	$1.50729 \cdot 10^{-1}$
10^4	$[1.33791, 1.33942]$	$1.50756 \cdot 10^{-3}$
10^6	$[1.33866, 1.33868]$	$1.50758 \cdot 10^{-5}$

Listing 5.8.   A simple-minded enclosure method using the PROFIL/BIAS package [PiBi]

```
1  #include <iostream>
2  #include "Interval.h"
3  #include "Functions.h"
4  using namespace std;
5  typedef INTERVAL (*pfcn)(INTERVAL);
6
7  INTERVAL integrate(pfcn f, INTERVAL x, int N) {
8    INTERVAL dX      = INTERVAL(0, Diam(x))/N;
9    INTERVAL localX = Inf(x);
10   INTERVAL quad    = 0;      /
11   for (int i = 0; i < N; i++) {
12     localX = Sup(localX) + dX;
13     if (Intersection(localX, localX, x))
14         quad += f(localX)*diam(localX);//Interval-valued diameter.
15   }
16   return quad;
17 }
18
19 INTERVAL integrand(INTERVAL X) { return sin(cos(exp(x))); }
20
21 int main()
22 {
23   INTERVAL X(atof(argv[1]), atof(argv[2]));
24   int      N(atoi(argv[3]));
25
26   cout << integrate(integrand, X, N) << endl;
27   return 0;
28 }
```

requires *one million* function evaluations to match the accuracy obtained with a mere 1,024 evaluations using the trapezoid method (5.12). This is to be expected, as the trapezoid method has a quadratic error term, whereas the error of our proposed enclosure method (5.15) is linear.

In order to improve the accuracy of our integral enclosures, we must generate tighter bounds on the integrand. This can be achieved by employing the techniques of automatic differentiation, described in Chapter 4. From now on, we will assume that the integrand f is n times continuously differentiable over the domain of integration: $f \in C^n(x)$. Then, using the notation $f_k(\tilde{x}) = f^k(\tilde{x})/k!$, we can

Taylor expand f around the midpoint $\tilde{x} = \text{mid}(\boldsymbol{x})$:

$$f(x) = \sum_{k=0}^{n-1} f_k(\tilde{x})(x - \tilde{x})^k + f_n(\zeta_x)(x - \tilde{x})^n. \tag{5.16}$$

Here, the point $\zeta_x \in \boldsymbol{x}$ is usually unknown. We can enclose the remainder term appearing in (5.16) by first computing $F_n(\boldsymbol{x})$, and then forming

$$\varepsilon_n = \text{mag}\big(F_n(\boldsymbol{x}) - f_n(\tilde{x})\big),$$

where we recall that $\text{mag}(\boldsymbol{a}) = \max\{|\underline{a}|, |\overline{a}|\}$. This produces the enclosure of the integrand

$$f(x) \in \sum_{k=0}^{n} f_k(\tilde{x})(x - \tilde{x})^k + [-\varepsilon_n, \varepsilon_n]|x - \tilde{x}|^n, \tag{5.17}$$

valid for all $x \in \boldsymbol{x}$. We are now prepared to compute the integral itself:

$$\int_{\tilde{x}-r}^{\tilde{x}+r} f(x)dx \in \int_{\tilde{x}-r}^{\tilde{x}+r} \left(\sum_{k=0}^{n} f_k(\tilde{x})(x - \tilde{x})^k + [-\varepsilon_n, \varepsilon_n]|x - \tilde{x}|^n \right) dx$$

$$= \sum_{k=0}^{n} f_k(\tilde{x}) \int_{-r}^{r} x^k dx + [-\varepsilon_n, \varepsilon_n] \int_{-r}^{r} |x|^n dx.$$

Now something interesting happens. Since we are integrating monomials over a domain centered at the origin, there is a lot of cancellation: all odd-numbered terms will evaluate to zero. Continuing our calculations, we have

$$\int_{\tilde{x}-r}^{\tilde{x}+r} f(x)dx \in \sum_{k=0}^{n} f_k(\tilde{x}) \int_{-r}^{r} x^k dx + [-\varepsilon_n, \varepsilon_n] \int_{-r}^{r} |x|^n dx$$

$$= \sum_{k=0}^{\lfloor n/2 \rfloor} f_{2k}(\tilde{x}) \int_{-r}^{r} x^{2k} dx + [-\varepsilon_n, \varepsilon_n] \int_{-r}^{r} |x|^n dx$$

$$= 2 \left(\sum_{k=0}^{\lfloor n/2 \rfloor} f_{2k}(\tilde{x}) \frac{r^{2k+1}}{2k+1} + [-\varepsilon_n, \varepsilon_n] \frac{r^{n+1}}{n+1} \right).$$

For large domains of integration, we can form a suitably fine partition $a = x_0 \le x_1 \le \cdots \le x_N = b$, and form the enclosure

$$\int_a^b f(x)dx = \sum_{i=1}^{N} \int_{x_{i-1}}^{x_i} f(x)dx = \sum_{i=1}^{N} \int_{\tilde{x}_i-r_i}^{\tilde{x}_i+r_i} f(x)dx$$

$$\in 2 \sum_{i=1}^{N} \left(\sum_{k=0}^{\lfloor n/2 \rfloor} f_{2k}(\tilde{x}_i) \frac{r_i^{2k+1}}{2k+1} + [-\varepsilon_{n,i}, \varepsilon_{n,i}] \frac{r_i^{n+1}}{n+1} \right). \tag{5.18}$$

In Listing 5.9, we present a simple C++ implementation of algorithm (5.18) that employs a uniform partition, that is, $x_i = a + ih$, where $h = (b - a)/N$ is the step-size. In Table 5.3, we present the outcome of some computations using the same piece of code, with Taylor expansions of degree six.

Listing 5.9.   An implementation of the Taylor-based enclosure method (5.18) using the
               PROFIL/BIAS package [PiBi] and a `taylor` class

```
 1  #include <iostream>
 2  #include "taylor.h"
 3  using namespace std;
 4  typedef taylor (*pfcn)(const taylor &);
 5
 6  INTERVAL riemannTerm(pfcn f, INTERVAL x, int Deg) {
 7    INTERVAL mid = INTERVAL(Mid(x));
 8    INTERVAL rad = diam(x)/2;    // Interval-valued diameter.
 9    taylor   fx  = f(variable(mid, Deg));
10    INTERVAL sum = fx[0]*rad;
11    for (int k = 2; k <= Deg; k += 2)
12      sum += fx[k]*Power(rad, k + 1)/(k + 1);
13    taylor Fx  = f(variable(x, Deg));
14    double eps = Abs(Fx[Deg] - fx[Deg]);
15    sum += INTERVAL(-eps, +eps)*Power(rad, Deg + 1)/(Deg + 1);
16    return 2*sum;
17  }
18
19  INTERVAL integrate(pfcn f, INTERVAL x, int Deg, int N){
20    INTERVAL dX      = INTERVAL(0, Diam(x))/N;
21    INTERVAL localX = Inf(x);
22    INTERVAL quad    = 0;
23    for (int i = 0; i < N; i++) {
24      localX = Sup(localX) + dX;
25     if (Intersection(localX, localX, x))
26        quad += riemannTerm(f, localX, Deg);
27    }
28    return quad;
29  }
30
31  taylor integrand(const taylor &x) { return sin(cos(exp(x))); }
32
33  int main(int argc, char * argv[])
34  {
35    INTERVAL   X(atof(argv[1]), atof(argv[2]));
36    int        Deg(atoi(argv[3]));
37    int        N(atoi(argv[4]));
38
39    cout << integrate(integrand, X, Deg, N);
40    return 0;
41  }
```

5.3.2 Adaptive Integration

Of course, it may be wasteful to split the domain of integration into a uniform grid.
A subinterval x_i that produces a very narrow Riemann term

$$T_n(f, x_i) = 2 \left(\sum_{k=0}^{\lfloor n/2 \rfloor} f_{2k}(\tilde{x}_i) \frac{r_i^{2k+1}}{2k+1} + [-\varepsilon_{n,i}, \varepsilon_{n,i}] \frac{r_i^{n+1}}{n+1} \right)$$

Table 5.3 Integral enclosures of $\int_{-2}^{2} \sin{(\cos{(e^x)})}dx$.

N	$E_{-2}^{2}(f, 6, N)$	$w(E_{-2}^{2}(f, 6, N))$
9	[0.86325178469, 1.81128961988]	$9.4804 \cdot 10^{-1}$
12	[1.28416304745, 1.39316025451]	$1.0900 \cdot 10^{-1}$
21	[1.33783795371, 1.33950680633]	$1.6689 \cdot 10^{-3}$
75	[1.33866863493, 1.33866878008]	$1.4514 \cdot 10^{-7}$

Table 5.4 Adaptive integral enclosures of $\int_{-2}^{2} \sin{(\cos{(e^x)})}dx$.

TOL	$A_{-2}^{2}(f, 6, \text{TOL})$	$w(A_{-2}^{2}(f, 6, \text{TOL}))$	N_{TOL}
10^{-1}	[1.33229594606, 1.34500942603]	$1.2713 \cdot 10^{-2}$	9
10^{-2}	[1.33822575109, 1.33911045235]	$8.8470 \cdot 10^{-4}$	12
10^{-4}	[1.33866170207, 1.33867571626]	$1.4014 \cdot 10^{-5}$	21
10^{-8}	[1.33866870618, 1.33866870862]	$2.4304 \cdot 10^{-9}$	75

need not be further decomposed. Instead, all efforts should be concentrated where the terms are relatively wide.

An adaptive integration scheme can easily be attained by recursively bisecting the original domain into a collection of subintervals $\{x_i\}_{i=1}^{N_{\text{TOL}}}$ whose corresponding Riemann terms meet the tolerance requirement:

$$w\big(T_n(f, x_i)\big) \leq \text{TOL}(x_i) \stackrel{\text{def}}{=} \text{TOL}([a, b])\frac{w(x_i)}{b - a}.$$

It is clear that the recursive subdivision must terminate after a finite number of steps. By (5.18) it follows that $w\big(T_n(f, x_i)\big) = \mathcal{O}(w(x_i)^{n+1})$. If the integrand f is of class C^{n+1}, then $f^{(n)}$ is of class C^1, and by Theorem 3.10, there exist positive real numbers K_i (that are uniformly bounded) such that $\varepsilon_{n,i} \leq K_i w(x_i)$. Therefore, we actually have

$$w\big(T_n(f, x_i)\big) = \mathcal{O}(w(x_i)^{n+2}),$$

which means that the tolerance requirement will be satisfied after a finite number of subdivisions, even for the case $n = 0$, corresponding to the naive enclosure method (5.14).

In Listing 5.10, we present a straightforward C++ implementation of the adaptive scheme just outlined.

Example 5.3.3 *Continuing from Example 5.3.2, let us now use the adaptive integration scheme described above. We will compute the interval enclosures using the tolerances* TOL $= 10^{-1}, 10^{-2}, 10^{-4},$ *and* 10^{-8}. *The order of the Taylor expansions is set to six.*

We present the result in Table 5.4. Here we have also listed the number of subintervals N_{TOL} *produced by the adaptive scheme. Comparing with Table 5.3, we see the advantage of adaptively decomposing the integration domain. Also see*

Listing 5.10. An adaptive version of algorithm (5.18) using the PROFIL/BIAS [PiBi] package and a taylor class

```
1  #include <iostream>
2  #include "taylor.h"
3  using namespace std;
4  typedef taylor (*pfcn)(const taylor &);
5
6  INTERVAL riemannTerm(pfcn f, INTERVAL X, int Deg) {
7    INTERVAL mid = INTERVAL(Mid(X));
8    INTERVAL rad = diam(X)/2; // Interval-valued diameter.
9    taylor   fx  = f(variable(mid, Deg));
10   INTERVAL sum = fx[0]*rad;
11   for (int k = 2; k <= Deg; k += 2)
12     sum += fx[k]*Power(rad, k + 1)/(k + 1);
13   taylor Fx  = f(variable(X, Deg));
14   double eps = Abs(Fx[Deg] - fx[Deg]);
15   sum += INTERVAL(-eps,+eps)*Power(rad, Deg + 1)/(Deg + 1);
16   return 2*sum;
17 }
18
19 INTERVAL integrate(pfcn f, INTERVAL X, int Deg, double Tol) {
20   INTERVAL sum = riemannTerm(f, X, Deg);
21   if (Diam(sum) <= Tol)
22     return sum;
23   else
24     return integrate(f, INTERVAL(Inf(X), Mid(X)), Deg, Tol/2) + \
25             integrate(f, INTERVAL(Mid(X), Sup(X)), Deg, Tol/2);
26 }
27
28 taylor integrand(const taylor &x) { return sin(cos(exp(x))); }
29
30 int main(int argc, char * argv[])
31 {
32   INTERVAL   X(atof(argv[1]), atof(argv[2]));
33   int        Deg(atoi(argv[3]));
34   double     Tol(atof(argv[4]));
35
36   cout << integrate(integrand, X, Deg, Tol);
37   return 0;
38 }
```

Figure 5.10, illustrating the successive enclosures over the partition of the domain of integration.

Exercise 5.14. *Note that the final enclosure widths in Table 5.4 are quite a bit smaller than requested. Can you explain why? How could one fix this?*

We end this section with an example due to Siegfried M. Rump.[5]

---

[5] Private communication with the author.

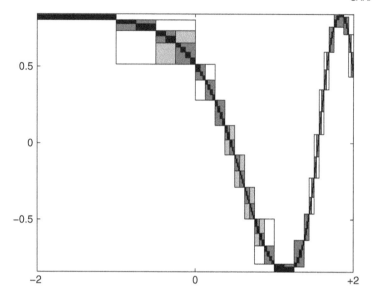

Figure 5.10 Refined enclosures of the integrand $f(x) = \sin(\cos(e^x))$.

Example 5.3.4 *Suppose we wish to compute the definite integral $\int_0^8 \sin(x + e^x)dx$. A regular* MATLAB *session could look as follows:*

```
% Define the intergrand and domain.
fcn_str = 'sin(x + exp(x))';
f = vectorize(inline(fcn_str));
a = 0; b = 8;
% Compute the integral using MATLAB's 'quad'.
>> q = quad(f,a,b)
q =
    0.251102722027180
```

Fine. No warnings are issued, and we are now convinced that we have the correct answer. The quad *method implements a low-order method using an adaptive recursive Simpson's rule. As a comparison, our adaptive validated integrator (with order 4, and tolerance 10^{-4}) produces the following output:*

```
$$ ./adQuad 0 8 4 1e-4
Partitions: 8542
CPU time   : 0.52 seconds
Integral   : 0.347[3863144222905,4140198005782]
```

The notation $0.347[3863144222905, 4140198005782]$ is to be interpreted as the interval $[0.3473863144222905, 0.3474140198005782]$. Thus, it turns out that the MATLAB *result was completely wrong. If we use Taylor expansions of order 20 and set the tolerance to 10^{-10}, we get*

```
$$ ./adQuad 0 8 20 1e-10
Partitions: 874
CPU time   : 0.45 seconds
Integral   : 0.3474001726[492276,652638]
```

Note that this actually took less time to compute. Using a high-order method reduces the number of partitions, thus saving time.

5.4 COMPUTER LAB IV

Problem 1. Write an interval bisection routine such as described in Section 5.1.1. Your program should prompt the user for the search domain x and a tolerance TOL. What are the results you get for the function $f(x) = \sin x - \cos x$ with $x = [-10, +10]$ and TOL $= 0.0001$?

Problem 2. Write an interval Newton routine. Your program should prompt the user for the search domain x. What are the results you get for the function $f(x) = \sin(\cos(x - 3))$ with the search domains $x_1 = [1, 2]$, $x_2 = [1.5, 2.5]$, and $x_3 = [2, 3]$?

Problem 3. Combine the ideas from the two previous problems and write a hybrid bisection/interval Newton routine. It should bisect the search domain into subintervals x_k until either diam$(x_k) \leq$ TOL or $0 \notin F'(x_k)$. In the latter case the subinterval should undergo an interval Newton search. Your program should prompt the user for the search domain x and a tolerance TOL. What are the results you get for the function $f(x) = \sin(\cos(x - 3))$ with the search domain $x = [-10, 10]$ and TOL $= 0.001$?

Problem 4. Write an interval optimizer, as described in Section 5.2. You may choose to use any or all of the midpoint, monotonicity, or convexity checks. Your program should prompt the user for the search domain x and a tolerance TOL. What are the results you get for the function

$$f(x) = x^2 - \frac{1}{2}e^{-(a(x - \frac{1}{2}))^2}$$

with $a = 10,000$, $x = [-10, +10]$, and TOL $= 10^{-10}$?

Problem 5. Write a program that computes rigorous enclosures of the standard quadrature methods by computing the approximations (5.12) in interval arithmetic and by enclosing the remainder terms (5.13) using Taylor series arithmetic (from Problem 4 of Computer Lab III).

Problem 6. Write a simple-minded interval integrator (as described in Section 5.3.1) and redo Example 5.3.1

Problem 7. Write an adaptive, Taylor-based, interval integrator (as described in Section 5.3.2) and redo Example 5.3.3

Chapter Six

Ordinary Differential Equations

IN THIS CHAPTER, we will describe various procedures for enclosing the solution to an ordinary differential equation (ODE).

6.1 A GENTLE MATHEMATICAL INTRODUCTION

In the real-valued setting, a general initial-value problem (IVP) can be formulated as follows: find a differentiable function $x \colon [0, T] \to \mathbb{R}$ that satisfies the following conditions:

$$
\begin{cases}
\dot{x}(t) = f(x(t), t) \\
x(0) = x_0.
\end{cases}
\tag{6.1}
$$

Here we use the notation $\dot{x} = \frac{d}{dt}x$. The function $f \colon D \times [0, T] \to \mathbb{R}$ is called a *vector field* and is assumed to satisfy a Lipschitz condition in its first variable, that is, there is a positive constant K such that, for all $t \in [0, T]$ and for all $x_1, x_2 \in D$, we have

$$
|f(x_1, t) - f(x_2, t)| \leq K |x_1 - x_2|.
$$

With this restriction on f, it is possible to prove that there exists a unique solution to (6.1) for every initial value $x_0 \in D$. The path defined by such a solution curve $x(t)$ is sometimes called an *orbit* or a *trajectory*. Given an initial condition $x_0 \in D$, it is not clear how long its corresponding trajectory exists, that is, we do not know the largest value of t for which $x(t)$ is well-defined. If we take the initial value x_0 very close to the boundary of the domain D, then its trajectory may leave the D in a very short time, after which we really have no control of the solution. On the other hand, if we restrict the initial values to a proper sub-domain $D' \subset D$, it is possible to obtain a lower bound T' on how long the solution exists (and is unique).

Exercise 6.1. *Given the IVP (6.1), where f has the Lipschitz constant K, find a minimal time of existence T' in terms of D and D'. You may assume that both domains are compact intervals.*

It is sometimes desirable to describe the solution to (6.1) as a *flow*. A flow is simply a solution where the dependence on the initial condition x_0 has been made explicit and is denoted $\varphi(x, t)$. The flow is defined by

$$
\begin{cases}
\dfrac{\partial}{\partial t}\varphi(x, t) = f(\varphi(x, t), t) \\
\varphi(x, 0) = x.
\end{cases}
\tag{6.2}
$$

Thus the trajectory passing through the point x at time $t = 0$ will be at position $\varphi(x, t)$ at time t.

Example 6.1.1 *Consider the ODE* $\dot{x} = -tx$. *The solution with initial condition* $x(0) = x_0$ *is given by* $x(t) = x_0 e^{-t^2/2}$ *(verify this). The corresponding flow is* $\varphi(x, t) = x e^{-t^2/2}$. *Note that the solutions are well-defined for all* $t \in \mathbb{R}$.

Now consider the ODE $\dot{x} = x^2$. *The solution with initial condition* $x(0) = x_0 \neq 0$ *is given by* $x(t) = (1/x_0 - t)^{-1}$ *(verify this). The corresponding flow is* $\varphi(x, t) = (1/x - t)^{-1}$. *This time, the solutions are well-defined only for* $t \in (-\infty, 1/x_0)$. *If* $x_0 = 0$, *then* $x(t) = \varphi(0, t) = 0$ *for all* $t \in \mathbb{R}$.

Our aim is to find an interval extension Ψ of the real-valued flow φ such that, for all $\boldsymbol{t} \subseteq [0, T']$ and $\boldsymbol{x} \subseteq D'$, we have the enclosure

$$R(\varphi; \boldsymbol{x} \times \boldsymbol{t}) \subseteq \Psi(\boldsymbol{x}, \boldsymbol{t}). \tag{6.3}$$

In particular, we should have $x_0 \in \boldsymbol{x}_0, t \in [0, T'] \Rightarrow \varphi(x_0, t) \in \Psi(x_0, t)$.

6.2 SIMPLE ENCLOSURE METHODS

To begin with, we will restrict our attention to the *autonomous* case, in which the vector field f does not explicitly depend on the variable t. The general IVP can then be formulated as

$$\begin{cases} \dot{x}(t) = f(x(t)), & t \in [0, T'] \\ x(0) = x_0. \end{cases} \tag{6.4}$$

We will use the fact that (6.4) combined with (6.2) can be expressed as the integral equation

$$\varphi(x_0, t) = x_0 + \int_0^t f(\varphi(x_0, s))ds. \tag{6.5}$$

As such, the solution $x(t)$ to (6.4) can be viewed as a fixed point to the integral operator

$$\mathcal{P}_{f,x_0}(\chi)(t) = x_0 + \int_0^t f(\chi(s))ds \tag{6.6}$$

acting on the space of continuously differentiable functions. Let us now *assume* that we have a continuous[1] interval-valued enclosure $\Psi(t)$ of the unique solution to (6.4). This could be produced by adding some width to an approximate solution obtained by a standard numerical solver. By inclusion isotonicity (2.5), it follows that the solution is enclosed also by the following interval-valued integral:

$$x(t) \in x_0 + \int_0^t F(\Psi(s))ds, \tag{6.7}$$

where F is an interval extension of the vector field f. This integral is defined by forming increasingly finer interval Riemann sums and taking the limit.[2] The

---

[1]By this, we mean that the two boundary graphs of $\Psi(t)$ are continuous.
[2]Taking the limit is not strictly necessary in this situation, as every finite interval Riemann sum contains the true integral.

question is whether we can use (6.7) to find a better enclosure than Ψ. To this end, we consider the sequence:

$$
\begin{aligned}
\Psi^{(0)}(t) &= \Psi(t), \\
\Psi^{(k+1)}(t) &= x_0 + \int_0^t F(\Psi^{(k)}(s))ds.
\end{aligned}
\tag{6.8}
$$

Note that each $\Psi^{(k)}(t)$ also encloses the solution $x(t)$. Using the Lipschitz property of f, it follows that $\mathrm{rad}\big(R(f;\mathbf{x})\big) \leq K\mathrm{rad}(\mathbf{x})$ for all $\mathbf{x} \subseteq D$. Thus, assuming that F also is Lipschitz[3] on D, we have $\mathrm{rad}(F(\mathbf{x})) \leq K'\mathrm{rad}(\mathbf{x})$ for some positive K'. Therefore, we also have

$$
\mathrm{rad}\big(\Psi^{(k+1)}(t)\big) \leq \mathrm{rad}\left(\int_0^t F(\Psi^{(k)}(s))ds\right) \leq \int_0^t \mathrm{rad}\big(F(\Psi^{(k)}(s))\big)ds
$$

$$
\leq \int_0^t K'\mathrm{rad}\big(\Psi^{(k)}(s)\big)ds \leq K't \max\big\{\mathrm{rad}\big(\Psi^{(k)}(s)\big): s \in [0,t]\big\}.
$$

Define $\rho_k = \max\{\mathrm{rad}(\Psi^{(k)}(s)): s \in [0,T']\}$; then we have $\rho_{k+1} \leq K't\rho_k \leq \cdots \leq (K't)^{k+1}\rho_0$. Thus, if T' is chosen small enough to guarantee that $K'T' < 1$, then the integral operator is a contraction (see Definition A.4).

Of course, it may seem hard to ensure that our starting assumption holds true: $x(t) \in \Psi(t)$ for all $t \in [0, T']$. Fortunately, all we must check in order to verify this is the following condition:

$$
x_0 + \int_0^t F(\Psi(s))ds \subseteq \Psi(t), \qquad t \in [0, T'].
\tag{6.9}
$$

If this condition holds, it follows by Brouwer's fixed point theorem (Theorem A.12) that the enclosure $\Psi(t)$ contains *at least* one fixed point of the integral operator (6.6). This means that it indeed encloses at least one solution to the IVP (6.4). By inclusion isotonicity, condition (6.9) also means that the sequence (6.8) is *nested*, that is, $\Psi^{(k+1)} \subset \Psi^{(k)}$.

Under these conditions, we can invoke Banach's fixed point theorem (Theorem A.14), where the complete metric space is given by the (closed) set of continuous functions χ whose graphs belong to Φ, and with the max-norm introduced above. It follows that the sequence of enclosures (6.8) will converge to the solution of the original IVP (6.4): $x(t) = \bigcap_{k=0}^{\infty} \Psi^{(k)}(t)$, for all $t \in [0, T']$.

In the absolutely simplest setting, we take all enclosures to be constant intervals, that is, we want to construct $\Psi^{(k)}(t) = \mathbf{z}^{(k)}$, $k \in \mathbb{N}$, where each $\mathbf{z}^{(k)}$ is an interval. We initialize the process by setting $\Psi^{(0)}(t) = \mathbf{z}^{(0)}$, where $\mathbf{z}^{(0)}$ contains the initial value x_0. Having done so, it is possible to compute the largest time T' such that for all $t \in [0, T']$, the new variant of the integral equation (6.8),

$$
\mathbf{z}^{(k+1)} = x_0 + \int_0^t F(\mathbf{z}^{(k)})ds = x_0 + [0, t] \times F(\mathbf{z}^{(k)}),
\tag{6.10}
$$

produces a nested sequence of intervals. It follows that a solution passing through x_0 does not leave any of the sets $\mathbf{z}^{(k)}$ before time T', that is, we can enclose the

---

[3] An interval-valued function F is Lipschitz if there is a positive K such that for all $\mathbf{x}$ we have $\mathrm{rad}\,(F(\mathbf{x})) \leq K\,\mathrm{rad}\,(\mathbf{x})$.

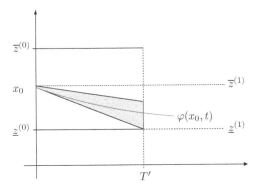

Figure 6.1 Computing successively tighter constant enclosures of the flow.

trajectory as follows:

$$R\big(\varphi(x_0, \cdot); [0, T']\big) \in \mathbf{z}^{(k)} \qquad k = 0, 1, \dots . \tag{6.11}$$

Just looking at a snapshot of the enclosures of the flows starting from x_0, that is, fixing a time $t \in [0, T']$, we have the possibly much tighter enclosures

$$\varphi(x_0, t) \in x_0 + t F(\mathbf{z}^{(k)}), \qquad k = 0, 1, \dots . \tag{6.12}$$

When we vary the parameter t, each of these enclosures forms a cone $C^{(k)}(t)$ extending from x_0. The rectangular hull of each cone $C^{(k)}(T')$ is contained in $\mathbf{z}^{(k)}$ (see Figure 6.1).

Seeing that the widths of the solution enclosures $C^{(k)}(t)$ are strictly increasing with time, it is fair to wonder whether we can ever obtain realistic results in a situation where, for example, all solutions are contracted toward a single trajectory. This is exactly what happens for solutions of $\dot{x} = -tx$, which we solved in Example 6.1.1 Independently of the initial value x_0, all trajectories approach the constant solution $x(t) = 0$ at an exponential rate. In particular, if we follow an entire interval of initial conditions $\mathbf{x}_0$ along the flow, its image is contracted as t increases. In this specific case we have an explicit formula for the image: $R\big(\varphi(\cdot, t); \mathbf{x}_0\big) = \mathbf{x}_0 e^{-t^2/2}$.

A successful way around this discouraging fact is to consider the endpoints of $\mathbf{x}_0$ separately. By the uniqueness property of the solutions, trajectories cannot cross, and therefore the two boundary orbits enclose the flow of the initial interval. Thus we know that for all $t \in [0, T']$ and $x \in \mathbf{x}_0$, we have the enclosure

$$\varphi(\underline{x}_0, t) \le \varphi(x, t) \le \varphi(\overline{x}_0, t).$$

Therefore, it suffices to compute enclosures of the two trajectories starting from the thin sets $\underline{x}_0$ and $\overline{x}_0$, respectively. A typical situation is illustrated in Figure 6.2. Here we see that although each enclosure is strictly increasing, their combined bounds form a contracting set.

Also note that if at some point in time t_1 the enclosures have become too wide for our liking, we simply merge the two enclosures to form a new interval

$$\mathbf{x}_1 = \Psi(\underline{x}_0, t_1) \sqcup \Psi(\overline{x}_0, t_1) \stackrel{\text{def}}{=} \tilde{\Psi}(\mathbf{x}_0, t_1).$$

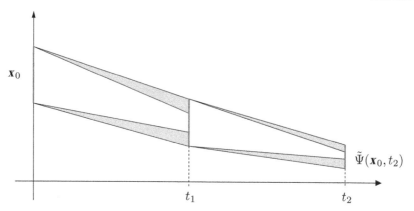

Figure 6.2 Obtaining global contraction with expanding local enclosures.

The entire procedure is then repeated starting with $\mathbf{x}_1$ instead of $\mathbf{x}_0$, producing

$$\mathbf{x}_2 = \Psi(\underline{x}_1, t_2 - t_1) \sqcup \Psi(\overline{x}_1, t_2 - t_1) \overset{\text{def}}{=} \tilde{\Psi}(\mathbf{x}_0, t_2)$$

and so on.

It should be pointed out that this approach only works in the one-dimensional setting. In higher dimensions, we cannot enclose the flow of an initial box by a finite number of trajectories, unless the solutions satisfy some monotonicity properties (see [Tu02]).

Given an initial set $\mathbf{x}_0$ and a positive time $T < T'$, we can obtain arbitrarily tight enclosures of the exact range $R(\varphi(\cdot, T); \mathbf{x}_0)$. This is achieved by splitting the interval $[0, T]$ into several subintervals $0 = t_0 < t_1 < t_2 < \cdots < t_n = T$ and performing the endpoint procedure described above on each subinterval $[t_i, t_{i+1}]$. The number of nodes t_i is determined by the desired accuracy (see Figure 6.3).

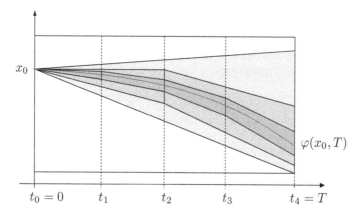

Figure 6.3 Increasingly tight enclosures of the trajectory $\varphi(x_0, t)$.

Exercise 6.2. *Write a program implementing the ideas from this section. Once a crude enclosure has been established, there are many different ways one can improve the enclosures (e.g., using centered forms, checking for monotonicity, etc.). Add some of your own improvements to the code.*

Exercise 6.3. *What changes must be made to allow for non-autonomous differential equations* $\dot{x} = f(x, t)$?

The methods presented in this section are *first-order* methods. As a consequence, the condition $K't < 1$ heavily restricts the maximal time steps allowed. Higher-order methods (e.g., Taylor series methods of a higher degree) partially overcome this problem, but the area is still an active field of research (see, e.g., [Mo66], [BM98], or [NJ01]).

6.3 HIGH-ORDER METHODS

One of the most powerful features of automatic differentiation is that it provides us with an elegant and effective means of computing numerical solutions to initial-value problems of the type (6.1). Considering only the autonomous setting for now, let $x(t)$ denote the solution, that is, $\dot{x}(t) = f(x(t))$, with initial condition $x(t_0) = x_0$.

Expanding $x(t)$ as a Taylor series centered at t_0, we have

$$x(t) = x_0 + x_1(t - t_0) + \cdots + x_k(t - t_0)^k + \cdots,$$

and by taking the formal derivative of this power series, we find that

$$\dot{x}(t) = x_1 + 2x_1(t - t_0) + \cdots + (k + 1)x_{k+1}(t - t_0)^k + \cdots.$$

Writing $f(x(t)) = (f \circ x)(t)$, we can express the Taylor series for the right-hand side of the differential equation, too:

$$(f \circ x)(t) = (f \circ x)_0 + (f \circ x)_1(t - t_0) + \cdots + (f \circ x)_k(t - t_0)^k + \cdots.$$

Now, a direct comparison of the Taylor expansions for $\dot{x}$ and $(f \circ x)$ gives a formal expression for the Taylor coefficients of the solution x:

$$x_{k+1} = \frac{(f \circ x)_k}{k + 1} \quad (k > 0). \tag{6.13}$$

Note that this is a recursive formula; computing the value of $(f \circ x)_k$ only requires coefficients of x up to order k. The initial value x_0 provides the seed for the recursion.

For efficiency, the value of $(f \circ x)_k$ can be obtained by the techniques presented in Chapter 4. By using automatic differentiation, we can thus numerically solve the IVP (6.1) by a Taylor method of any desired order. This should be viewed in stark contrast to, for example, the Runge-Kutta methods, each of which has a fixed order of convergence. The ability to change the order of the Taylor series is as important as varying the time-step in a traditional solver. We thus have a method for which we can vary both the time-step and the order at run-time and at a very low computational cost. This makes the solver more robust and accurate than traditional fixed-order variants. A thorough discussion of these issues is presented in [JZ05], with a nice application to the three-body problem.

Exercise 6.4. *The code presented in Listing 6.1 solves the ODE* $\dot{x} = \sin(1 + \cos x)$ *for any given initial value* x_0. *Modify the code to handle non-autonomous differential equations.*

6.4 RIGOROUS HIGH-ORDER EXAMPLES

In what follows, we will incorporate remainder bounds into the Taylor series method, thus making it rigorous. By the Mean Value Theorem, we have

$$\varphi(x, t + h) = \varphi_0(x, t) + \varphi_1(x, t)h + \cdots + \varphi_n(x, t)h^n + \varphi_{n+1}(x, s)h^{n+1},$$

for some $s \in [t, t + h]$. Note that the Taylor coefficients $\varphi_k = \varphi_k(x, t)$ of the flow φ can be computed according to (6.13). Combining the ideas presented in Section 6.2 with the use of Taylor series enables us to enclose a trajectory of a scalar ODE with a high-order method.

Assume that we are at stage k, that is, we have computed $\mathbf{x}_k$, which is a rigorous enclosure of $\varphi(\mathbf{x}_0, t_k)$. Now we want to proceed to step $k + 1$. We will, once again, use the fact that we are working in one phase variable and treat the endpoints of $\mathbf{x}_k$ separately, similarly to Figure 6.2.

Using, for example, the first-order method (6.10), we start by computing a feasible flow time $h_k = t_{k+1} - t_k$ and the associated crude enclosures $\hat{\mathbf{z}}_k$ and $\check{\mathbf{z}}_k$ of the two endpoints' flows during this time. The flow time is obtained by taking the minimum of the the computed flow times for both endpoints of $\mathbf{x}_k$. Next, we compute

$$\hat{\mathbf{w}}_{k+1} = \varphi_0(\overline{x}_k, t_k) + \cdots + \varphi_n(\overline{x}_k, t_k)h_k^n + \varphi_{n+1}(\overline{x}_k, [t_k, t_{k+1}])h_k^{n+1},$$

$$\check{\mathbf{w}}_{k+1} = \varphi_0(\underline{x}_k, t_k) + \cdots + \varphi_n(\underline{x}_k, t_k)h_k^n + \varphi_{n+1}(\underline{x}_k, [t_k, t_{k+1}])h_k^{n+1},$$

and check that the enclosures $\hat{\mathbf{w}}_{k+1} \subseteq \hat{\mathbf{z}}_k$ and $\check{\mathbf{w}}_{k+1} \subseteq \check{\mathbf{z}}_k$ are valid. Note that all "thin" Taylor coefficients must be computed separately from the two final "wide" coefficients. The latter are computed using the first-order enclosures $\varphi(\overline{x}, [t_k, t_{k+1}]) \subseteq \hat{\mathbf{z}}_k$ and $\varphi(\underline{x}, [t_k, t_{k+1}]) \subseteq \check{\mathbf{z}}_k$, which make them of inferior quality compared to the coefficients computed over point domains. Nevertheless, these "wide" coefficients are scaled by the (hopefully very small) factor h_k^{n+1}. This allows for a tight enclosure of the image of each endpoint of $\mathbf{x}_k$ under the flow.

As a final step, we take the hull of both enclosures:

$$\mathbf{x}_{k+1} = \check{\mathbf{w}}_{k+1} \sqcup \hat{\mathbf{w}}_{k+1}.$$

This completes one full integration step.

We conclude this section by illustrating the described method applied to a small selection of initial-value problems. As described above, we will consider set-valued IVPs, that is, we will solve for sets of initial conditions.

Listing 6.1.   A Taylor series solver for autonomous initial-value problems

```
1  #include <iostream>
2  #include "taylor.h"
3  using namespace std;
4
5  taylor f(const taylor &x) {
6    return sin(1 + cos(x));
7  }
8
9  taylor taylorODE(double x0, int N) {
10   taylor phi = constant(x0, N);
11   for (int k = 0; k < N; k++) {
12     double newTerm = f(phi)[k] / (k + 1);
13     setCoeff(phi, k + 1, newTerm);
14   }
15   return phi;
16 }
17
18 double evaluate(const taylor &a, const double &t)
19 {
20   int    N   = orderOf(a);
21   double sum = a[N];
22   for (int i = N - 1; i >= 0; i--)
23     sum = sum*t + a[i];
24   return sum;
25 }
26
27 void solveODE(double x0, double t0, double tf, int Ng, int No) {
28   double dt = (tf - t0)/Ng;
29   double  x = x0;
30   cout << t0 << "  \t " << x0 << endl;
31   for (int i = 1; i <= Ng; i++) {
32     double  t = t0 + i*dt;
33     taylor Tx = taylorODE(x, No);
34     x = evaluate(Tx, dt);
35     cout << t << "  \t " << x << endl;
36   }
37 }
38
39 int main(int argc, char * argv[])
40 {
41   if (argc != 6) {
42     cerr << "Syntax: nrODE x0 t0 tf Ng No" << endl;
43     exit(0);
44   }
45   double x0 = atof(argv[1]); // Initial value.
46   double t0 = atof(argv[2]); // Initial time.
47   double tf = atof(argv[3]); // Final time.
48   int Ng = atoi(argv[4]);    // Number of grid points.
49   int No = atoi(argv[5]);    // Order of Taylor expansion.
50
51   solveODE(x0, t0, tf, Ng, No);
52
53   return 0;
54 }
```

Example 6.4.1 *As a first example, let us consider the non-autonomous IVP*

$$\begin{cases} \dot{x}(t) = -tx(t), & t \geq t_0 \\ x(t_0) = x_0 \in [-1, 1], \end{cases} \tag{6.14}$$

which has the exact solution

$$x(t) = x_0 e^{-(t^2 - t_0^2)/2}.$$

From this it is clear that $x \equiv 0$ is asymptotically stable. We present the enclosures to the problem (6.14) in Figure 6.4.

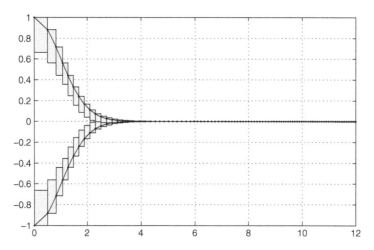

Figure 6.4 Third order interval solution to (6.14) with $\mathbf{x}_0 = [-1, 1]$, up to $t = 12$. The lower (upper) boxes are the verification boxes for the lower (upper) endpoints. The lines are linear interpolations of these endpoints. As seen, the interval solution contracts rapidly around $x = 0$; at time $t = 12$ the diameter of the computed interval solution is less than 10^{-17}.

Example 6.4.2 *The autonomous IVP*

$$\begin{cases} \dot{x}(t) = x(t)^2, & t \geq 0 \\ x(0) = x_0 \in [1, 1.25] \end{cases} \tag{6.15}$$

has the solution

$$x(t) = \frac{1}{1/x_0 - t}, \quad t \in [0, 1/x_0),$$

which blows up at time $t_c = 1/x_0$. We present the computed enclosures to the problem (6.15) in Figure 6.5.

The next two examples are point-valued, that is, the solution is a classical (non-interval) trajectory. Using a set-valued solver illustrates stability properties of the underlying differential equations.

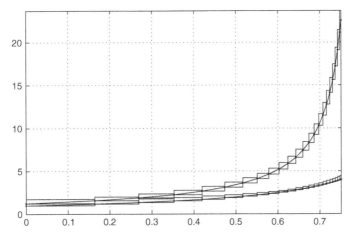

Figure 6.5 Third-order interval solution to (6.15) with $x_0 = [1, 1.25]$, up to $t = 0.75$. The diameters of the upper verification boxes grow rapidly as t approaches $t_c = 0.8$.

Example 6.4.3 *Consider the autonomous IVP*

$$\begin{cases} \dot{x}(t) = -x(t)^3, & t \geq 0 \\ x(0) = 1. \end{cases} \tag{6.16}$$

We present the computed enclosures to the problem (6.16) in Figure 6.6.

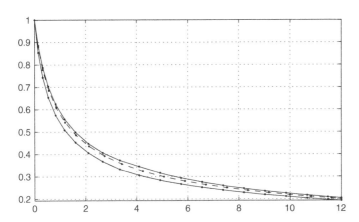

Figure 6.6 A first-order interval enclosure (outer pair) and a tighter fifth-order interval enclosure (inner pair) to (6.16). The verification boxes are not shown. The maximal diameter is 5×10^{-2} for the the first-order enclosure, and 2×10^{-4} for the fifth-order enclosure.

Example 6.4.4 *Consider the following autonomous IVP:*

$$\begin{cases} \dot{x}(t) = x(t)(x(t) - 1), & t \geq 0 \\ x(0) = 1, \end{cases} \tag{6.17}$$

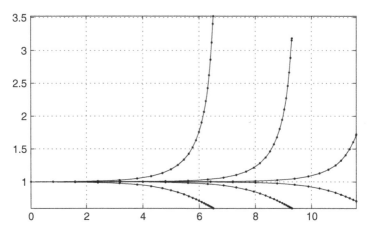

Figure 6.7 Increasingly tight enclosures to (6.17) using orders 3, 5, and 7.

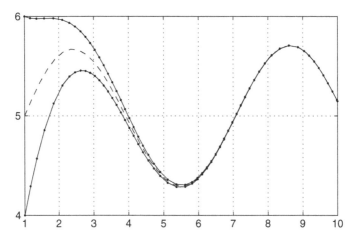

Figure 6.8 Computed enclosures of (6.18) with $t_0 = 1$, $x_0 = [4, 6]$. The dashed line is the MATLAB ode45 solution with initial condition $x_0 = 5$.

which has the solution $x(t) = 1$, $t \geq 0$. *We compute interval enclosures of (6.17) using Taylor series of order 3, 5, and 7. We stop the computation when the diameter of the enclosure grows too large. We present the enclosures to the problem (6.17) in Figure 6.7. Note that the third-order solution was stopped at $t = 6$, the fifth-order at $t = 9$, and the seventh-order at $t = 12$.*

We now switch back to set-valued problems.

Example 6.4.5 *Let us consider the following non-autonomous IVP:*

$$\begin{cases} \dot{x}(t) = 5 + \sin t - x(t), & t \geq 1 \\ x(1) = [4, 6]. \end{cases} \tag{6.18}$$

Table 6.1 Maximal diameter of different order Taylor series solutions to (6.19)

Taylor solver order	1	2	3	4
maximal diameter	1.0×10^{-2}	1.9×10^{-3}	7.1×10^{-4}	5.7×10^{-4}

For a third-order Taylor method, the diameter of the enclosure at $t = 10$ is 2×10^{-4}. For a sixth-order Taylor method, the diameter decreased only slightly. We present the enclosures to the problem (6.18) in Figure 6.8.

Example 6.4.6 *We end with the somewhat complicated non-autonomous IVP*

$$\begin{cases} \dot{x}(t) = f(x, t), \quad t \in [0, 10] \\ x(0) = [3 - \varepsilon_M, 3 + \varepsilon_M] \end{cases} \tag{6.19}$$

where

$$f(x, t) = \frac{e^{e^{-tx}} + \frac{x^3}{100} + \frac{x}{10} + 2 + 10 \cos x + 4 \sin t - \log x}{\frac{x^3}{50} + 4x^2 + 3x + 4 + \frac{\sin 1.5tx}{1000}(x + 1)^{0.75} + \frac{\cos 3.14t}{1000}},$$

and ε_M denotes the machine epsilon. We present the enclosures to the problem (6.19) in Figure 6.9. Table 6.1 shows the maximal diameter of the enclosures for different orders of the Taylor series solver.

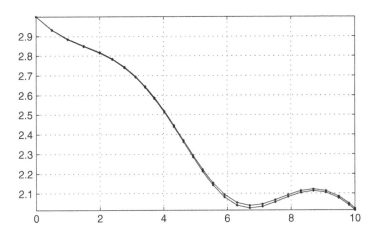

Figure 6.9 The first-order enclosure of (6.19).

Appendix A

Mathematical Foundations

IN THIS APPENDIX, we will explore the set of real numbers $\mathbb{R}$, which we will regard as a utopian model set on which we can perform perfect computations. When computing over more sparse sets, we will try to recover, or at least approximate, most of the properties valid over $\mathbb{R}$. We also present the important fixed point theorems, which are crucial for most of the interval methods described in the book.

A.1 THE RATIONAL NUMBERS

It is customary to define the real numbers in terms of the set of rational numbers, denoted by $\mathbb{Q}$. A rational number p/q can be thought of as an ordered couple of integers (p, q) with the restriction $q \neq 0$. Assuming that we fully understand the set of integers $\mathbb{Z}$, we may define the arithmetic properties of $\mathbb{Q}$:

$$p/q + p'/q' = (pq' + p'q)/qq' \quad \text{(addition)}$$
$$p/q - p'/q' = (pq' - p'q)/qq' \quad \text{(subtraction)}$$
$$p/q \times p'/q' = pp'/qq' \quad \text{(multiplication)}$$
$$p/q \div p'/q' = pq'/qp' \quad \text{(division)}$$

where we demand that $p' \neq 0$ when dividing.

We may identify the rational number p/q with any rational number p'/q' such that $pq' - p'q = 0$. Thus the representation of a rational number is not unique. It is common to use the notation 0 for any rational number p/q with $p = 0$ and 1 for any rational number p/q with $p = q$.

The rational numbers form a *field* under the operations $\{+, \times\}$. This means

1. 0 is the additive identity, and 1 is the multiplicative identity:

$$0 + x = x; \qquad 1 \times x = x;$$

2. every rational number has an additive inverse, and every non-zero rational number has a multiplicative inverse:

$$x + (-x) = 0; \qquad x \times (1/x) = 1;$$

3. addition and multiplication are both commutative and associative:

$$x + y = y + x;$$
$$(x + y) + z = x + (y + z);$$

$$x \times y = y \times x;$$

$$(x \times y) \times z = x \times (y \times z);$$

4. multiplication distributes over addition:

$$x \times (y + z) = x \times y + x \times z,$$

We can impose an *order* on the rational numbers by defining the sign of a rational number: we say that p/q is *positive* if p and q have the same sign; it is *negative* if p and q have different signs; it is *zero* if p is zero. Using this terminology, we define an ordering as follows: $x < y$ if $y - x$ is positive; $x \leq y$ if $y - x$ is positive or zero. Thus the set of rational numbers forms an *ordered field*.

In contrast to, for example, the integers, the rational numbers satisfy the Axiom of Archimedes, which states that for every positive rational number x, there exists an integer n such that $1/n < x$. This "axiom" is actually a theorem that can be proved for the rational numbers. One important consequence is that there is no smallest positive rational number. Regardless of how small we take x, we can always find another rational y such that $0 < y < x$. As we shall see, this property is also shared by the real numbers.

$$
\begin{array}{cccccc}
1/1 & 2/1 & \rightarrow & 3/1 & 4/1 & \rightarrow & 5/1 & \cdots \\
\downarrow & \nearrow & & \swarrow & \nearrow & & \swarrow \\
1/2 & 2/2 & & 3/2 & 4/2 & & 5/2 & \cdots \\
& \swarrow & & \nearrow & & \swarrow \\
1/3 & 2/3 & & 3/3 & 4/3 & & 5/3 & \cdots \\
\downarrow & \nearrow & & \swarrow \\
1/4 & 2/4 & & 3/4 & 4/4 & & 5/4 & \cdots \\
& \swarrow \\
1/5 & 2/5 & & 3/5 & 4/5 & & 5/5 & \cdots \\
\vdots & \vdots & & \vdots & \vdots & & \vdots & \ddots
\end{array}
$$

Figure A.1 A countable enumeration of the positive rational numbers.

Following Cantor, we say that two sets have the same *cardinality* if their elements can be put in a one-to-one correspondence with each other. This definition certainly makes sense when comparing finite sets, but its main strength becomes clear only with infinite sets. A set with the same cardinality as the integers is said to be *countable*. It usually comes as a surprise to novice students that $\mathbb{Q}$ is countable. Intuition leads us to believe that there are many more rational numbers than integers, whereas they in fact are in one-to-one correspondence.

THEOREM A.1. (Countability of $\mathbb{Q}$) *The set of rational numbers is countable.*

Proof. To prove countability, it suffices to show that we can label all rational numbers using the integers. To achieve this, we construct an infinite array of rational numbers (see Figure A.1).

It is clear that every rational number will appear at least once in this array. We label the elements by following the diagonals, moving from the left to the right, omitting all numbers that already have appeared. Thus the ten first elements are

$$\{1/1, 1/2, 2/1, 3/1, 1/3, 1/4, 2/3, 3/2, 4/1, 5/1\}.$$

Note that $2/2$ has been excluded from the numbering, since the equivalent number $1/1$ had already been encountered. This labeling gives a one-to-one correspondence between the positive integers and the positive rational numbers. The negative rational numbers (and zero) can be handled analogously. □

A.2 WHAT IS A REAL NUMBER?

As we have seen, the set of rational numbers is closed under the arithmetic operations $\{+, -, \times, \div\}$, with the exception of division by zero, which is undefined. In spite of this, this set is insufficient for our needs. Even the Greek mathematicians realized that the length of the hypotenuse of a right-angled, isosceles triangle with sides of unit length could not be represented as a rational number. A formal proof is given in the following theorem.

THEOREM A.2. *There is no rational number x such that $x^2 = 2$.*

Here we are using the common notation $x^n = x \times \cdots \times x$, with n factors. The theorem is surprising seeing that the rational numbers are so abundant. Indeed, given any two rational numbers x and y, there are infinitely many rational numbers between them. This is a direct consequence of the Axiom of Archimedes. In view of this fact, the rationals can approximate any number to any desired degree of accuracy. Nevertheless, by the theorem we are about to prove, it is clear that some numbers (most, actually) are not included in the set $\mathbb{Q}$. The proof is straightforward, provided that the reader is familiar with the concepts of prime factorization over the integers.

Proof. Either there is a rational number satisfying $x^2 = 2$ or there is not. Assume that there *is* such a number $x = p/q$. This implies that $x^2 = p^2/q^2 = 2$, which immediately gives a contradiction since both p^2 and q^2 have an even number of each prime appearing in their corresponding factorizations. More precisely, there are two odd integers K_p and K_q such that $p^2 = 2^{2M_p} K_p$ and $q^2 = 2^{2M_q} K_q$. Thus the number of twos appearing in the quotient p^2/q^2 is even (or zero). □

Note that the theorem (and its proof) works equally well with the number 2 replaced by any other prime number.

The reason why there are "gaps," such as the number $\sqrt{2}$, between the rational numbers can be explained by considering Cauchy sequences over $\mathbb{Q}$. In order to do this, we must first introduce a notion of distance. This is achieved by a *metric*.

DEFINITION A.3 ((Metric Space)). *A metric space (X, d) is a pair consisting of a set X and a function $d : X \times X \to \mathbb{R}$ satisfying*

1. *$d(x, y) \geq 0$ for all $x, y \in X$. Also, $d(x, y) = 0$ if and only if $x = y$;*
2. *$d(x, y) = d(y, x)$ for all $x, y \in X$;*
3. *$d(x, y) \leq d(x, z) + d(z, y)$ for all $x, y, z \in X$.*

This definition, however, presupposes that we understand what is meant by $\mathbb{R}$, which we actually have not yet defined. For now, we can replace $\mathbb{R}$ with $\mathbb{Q}$.

Example A.2.1 Let $\rho\colon \mathbb{Q} \times \mathbb{Q} \to \mathbb{Q}$ be defined by $\rho(x, y) = \text{sign}(x - y) \times (x - y)$. Then $(\mathbb{Q}, \rho)$ is a metric space. It is simply the set of rational numbers equipped with the Euclidean distance.

It is customary to use the notation $|x - y|$ to denote the Euclidean distance between two points x and y.

Example A.2.2 Let $\sigma\colon \mathbb{Q} \times \mathbb{Q} \to \mathbb{Q}$ be defined by $\sigma(x, y) = 0$ if $x = y$, and $\sigma(x, y) = 1$ if $x \neq y$. Then $(\mathbb{Q}, \sigma)$ is a metric space—it is called the discrete metric on $\mathbb{Q}$.

DEFINITION A.4 ((Cauchy Sequence)). A Cauchy sequence of rational numbers is a sequence $\{x_i\}_{i=1}^{\infty}$ with $x_i \in \mathbb{Q}$ such that given $\varepsilon > 0$ there exists $N > 0$ such that $|x_n - x_m| < \varepsilon$ for all $n, m \geq N$.

In other words, the *entire* tail $\{x_i\}_{i=N}^{\infty}$ of a Cauchy sequence can be confined to an increasingly small region as N tends to ∞. This implies that *all* elements of the tail are getting closer to each other and hence to their common limit. This is quite different from the requirement that all *consecutive* elements are to approach each other, that is, $\lim_{i \to \infty} |x_i - x_{i+1}| = 0$. In this latter case, the sequence may not even have a well-defined limit.

The major advantage of Cauchy sequences is that we can derive properties of the limit of the sequence without ever having to explicitly use it in our computations. All information is extracted exclusively from the elements of the sequence. For example, one can prove that a Cauchy sequence always converges to one unique number: its limit.

Returning to the problem with the missing numbers (the "gaps"), we can explain the strange phenomenon by the following observation: the limit of a rational Cauchy sequence need not be a rational number. In other words, $\mathbb{Q}$ is not closed under limits of Cauchy sequences. Clearly, we want to include numbers x such that $x^2 = 2$ in our set of real numbers. To achieve this, let us define $\mathbb{R}$ to consist of all rational numbers *and* all limits of rational Cauchy sequences. Thus the set $\mathbb{R}$ consists of all numbers that are, or can be approximated by, rational numbers.

The set of real numbers is, just like $\mathbb{Q}$, an ordered field. One simply has to check that the arithmetic operations over the reals preserve Cauchy sequences of rationals. The proofs are rather lengthy and are therefore omitted here. For a thorough treatment of the construction of the real numbers, see, for example, [St95]. One striking difference between $\mathbb{Q}$ and $\mathbb{R}$, on the other hand, is that of cardinality: there are *many* more real numbers than rationals.

THEOREM A.5. (Uncountability of $\mathbb{R}$) The set of real numbers is uncountable.

Proof. In fact, we will prove that the subset $(0, 1) \subset \mathbb{R}$ is uncountable. To achieve this, we will (falsely) assume that $(0, 1)$ is countable and then derive a contradiction. If $(0, 1)$ is countable, then we can produce a list of *all* of its elements, say,

$\{r_i\}_{i=1}^{\infty}$. We will now produce an element r of $(0, 1)$ that is *not* in the list $\{r_i\}_{i=1}^{\infty}$. This gives the contradiction and thus proves the theorem.

Consider all of the numbers r_i, $i = 1, 2, \ldots$ written as ordinary decimal numbers such that all expansions ending in an infinite sequence of 9s are rewritten as numbers ending in an infinite sequence of 0s. By this we mean that instead of writing, for example, $0.456999999\ldots$, we use the expression $0.457000000\ldots$. This ensures that every number has a unique decimal representation. Let us now construct the missing number r by specifying its decimal representation. If the i:th decimal of r_i is 5, we let the i:th decimal of r be 4; otherwise, we let it be 5. Clearly, r is a number in $(0, 1)$ whose decimal expansion only contains the numbers 4 and 5. By construction, we have also ensured that $r \neq r_i$ for all $i = 1, 2, \ldots$. Thus we have a contradiction, and $(0, 1)$ (and hence $\mathbb{R}$) is uncountable. □

Although $\mathbb{R}$ is uncountable (whereas $\mathbb{Q}$ is countable), the set of rational numbers can approximate any real number to any desired accuracy. This fact is made more precise as follows.

THEOREM A.6. (Density of $\mathbb{Q}$) *Given any real number x and a predefined error $\varepsilon > 0$, there exists a rational number y such that $|x - y| \leq \varepsilon$.*

Proof. Let $\{x_i\}_{i=1}^{\infty}$ be a rational Cauchy sequence with limit x. By definition, given $\varepsilon > 0$, there exists $N > 0$ such that $|x_n - x_m| < \varepsilon$ for all $n, m \geq N$. Let $y = x_N$. Then $|x_n - y| \leq \varepsilon$ for all $n \geq N$. In particular, the limit x also satisfies this inequality. □

From a computational point of view, this property is extremely important. The set of rational numbers is much smaller than $\mathbb{R}$, yet its elements are arbitrarily close to the real numbers. In some sense, this means that we are not losing too much information even if we choose to compute with a rational instead of a real number. For example, if we only need to know a numerical quantity up to its first ten decimals, it would be wasteful to compute the quantity with real numbers having infinite decimal expansions. In fact, in this case even the set of rational numbers is far too large a set for our needs.

Exercise A.7.

(a) Prove that $(\mathbb{R}^n, \rho)$ is a metric space if $\rho : \mathbb{R}^n \times \mathbb{R}^n \to \mathbb{R}$ is defined by $\rho(x, y) = \max\{|x_i - y_i| : i = 1, \ldots, n\}$.

(b) Prove that $(\mathbb{R}^n, \delta)$ is not a metric space if $\delta : \mathbb{R}^n \times \mathbb{R}^n \to \mathbb{R}$ is defined by $\delta(x, y) = \min\{|x_i - y_i| : i = 1, \ldots, n\}$.

A.3 COMPLETENESS

The most fundamental property of the real numbers is that of completeness. A nonempty set of numbers S is said to be *complete* if the limit of any Cauchy sequence in S belongs to S. As we just have seen, this is clearly not the case when $S = \mathbb{Q}$. But what about the case $S = \mathbb{R}$? Perhaps we will get a new set of numbers as limits of *real* Cauchy sequences. This may then go on forever and ever—clearly a non-desirable situation. Fortunately this is not the case: $\mathbb{R}$ is closed under limits of real Cauchy sequences, that is, the limit point of a real Cauchy sequence is a real number.

This terminology allows us to distinguish $\mathbb{Q}$ from $\mathbb{R}$. We have already seen that a sequence of rational numbers need not have a rational limit (e.g., the truncated decimal expansion of $\sqrt{2}$: $1, 1.4, 1.41, \ldots$), that is, $\mathbb{Q}$ is not closed under limits of Cauchy sequences. In this respect, the real numbers are very different.

THEOREM A.8. (Completeness of $\mathbb{R}$) *The limit of a real Cauchy sequence is a real number.*

Proof. Let $\{x_i\}_{i=1}^{\infty}$ be a Cauchy sequence of real numbers. We want to prove that it has a real number as its limit, that is, we want to construct a rational Cauchy sequence $\{\xi_i\}_{i=1}^{\infty}$ with limit ϱ and then prove that $\lim_{i \to \infty} x_i = \varrho$.

Let us choose ξ_i to be a rational number satisfying $|\xi_i - x_i| \leq 1/i$. This can be achieved since the rational numbers are dense in $\mathbb{R}$. Having done so, let us prove that $\{\xi_i\}_{i=1}^{\infty}$ indeed is a Cauchy sequence. Given $\varepsilon > 0$, we take $N > 0$ such that $|x_m - x_n| \leq \varepsilon/2$ for all $m, n \geq N$. This is possible since $\{x_i\}_{i=1}^{\infty}$ is a Cauchy sequence. By the triangle inequality, we then have $|\xi_m - \xi_n| \leq |\xi_m - x_m| + |x_m - x_n| + |x_n - \xi_n| \leq 1/m + \varepsilon/2 + 1/n$, which can be made less than ε by taking N (and thus m and n) sufficiently large. Hence $\{\xi_i\}_{i=1}^{\infty}$ is a rational Cauchy sequence, and so its limit is a real number, say ϱ.

Now, let us conclude by showing that $\lim_{i \to \infty} x_i = \varrho$. Using the triangle inequality once more, we have $|\varrho - x_i| \leq |\varrho - \xi_i| + |\xi_i - x_i| \leq |\varrho - \xi_i| + 1/i \leq \varepsilon/2 + 1/i$ for $i \geq N$. Again, this can be made less than ε by taking N (and thus i) sufficiently large. Hence $\lim_{i \to \infty} x_i = \varrho$. $\qquad \square$

Exercise A.9. *Let $S = (0, 1)$. Prove that S equipped with the Euclidean metric $d(x, y) = |x - y|$ is not a complete metric space.*

Another important property of the real numbers is usually stated in terms of the existence of a *least upper bound*.

DEFINITION A.10. (Least Upper Bound) *Given a subset S of the real numbers, we call a number γ an upper bound for S if for every $s \in S$ we have $s \leq \gamma$. Such a number γ is called a least upper bound for S (denoted $\sup S$) if no other upper bound for S is smaller than γ.*

Note that every finite set S trivially has a least upper bound: we simply choose the largest element of S as the least upper bound. If, however, S is an infinite set, things are not so obvious. In many cases, the least upper bound of S will not even be a member of S but rather the element of the ambient space ($\mathbb{R}$) that "puts the lid" on S.

THEOREM A.11. (Least Upper Bound) *Every non-empty subset S of the real numbers that has an upper bound also has a least upper bound.*

Proof. According to the assumptions, the set S has at least one element s_1 in addition to an upper bound γ_1. By definition, we must have $s_1 \leq \gamma_1$. If in fact

$s_1 = \gamma_1$, we are done: sup $S = \gamma_1$. If this is not the case, we consider the midpoint $\eta_1 = (s_1 + \gamma_1)/2$. If η_1 is a new (smaller) upper bound for S we set $\gamma_2 = \eta_1$ and $s_2 = s_1$. If η_1 is not an upper bound for S, there must be an element $s_2 \in S$ with $\eta_1 < s_2$. If this is the case, we set $\gamma_2 = \gamma_1$. In both cases, we end up with a new pair of numbers s_2 and γ_2, where $s_2 \in S$ and γ_2 is an upper bound for S. By construction, we also have $|s_2 - \gamma_2| \le \frac{1}{2}|s_1 - \gamma_1|$.

By repeating this procedure, we either terminate at a finite stage k with sup $S = \gamma_k$ or we generate a non-decreasing sequence $\{s_i\}_{i=1}^{\infty}$ of elements of S and a non-increasing sequence $\{\gamma_i\}_{i=1}^{\infty}$ of upper bounds of S, with $|s_k - \gamma_k| \le \frac{1}{2^{k-1}}|s_1 - \gamma_1|$. It follows that both generated sequences are Cauchy sequences and that they have the same limit, say, ϱ. Since, for any $s \in S$ we have $s \le \gamma_i$, it follows that $s \le \lim_{i \to \infty} \gamma_i = \varrho$. Thus ϱ is an upper bound for S. Likewise, seeing that $\lim_{i \to \infty} s_i = \varrho$, it follows that there cannot exist a smaller upper bound for S. Hence sup $S = \varrho$. $\square$

A.4 FIXED POINT THEOREMS

In 1910, the Dutch mathematician L.E.J. Brouwer proved that a continuous map taking the closed unit ball in $\mathbb{R}^n$ into itself must leave at least one point fixed (see [Br10]). By the closed unit ball in $\mathbb{R}^n$, we mean the following set:

$$\mathbb{D}^n = \{x \in \mathbb{R}^n : x_1^2 + \cdots + x_n^2 \le 1\}.$$

THEOREM A.12. (**Brouwer's Fixed-Point Theorem**) *Every continuous map*

$$f : \mathbb{D}^n \to \mathbb{D}^n$$

has at least one fixed point.

A nice proof is presented in [Mi65]. There the strategy is to first establish the theorem for smooth maps. Next, an approximation argument is used to pass to the continuous case. Being a purely topological statement, the theorem is also valid for any continuous deformation of $\mathbb{D}^n$, that is, an n-dimensional box. Brouwer's fixed point theorem is the foundation of the interval Newton method (see Theorem 5.5) in higher dimensions.

We next present a fixed point theorem that was proved by S. Banach in his 1922 Krakow dissertation (see [Ba22]). We first introduce the notion of *contractions*.

DEFINITION A.13 (**Contraction**). *A mapping T from a complete metric space (X, d) into itself is called a contraction if there exists a $\kappa \in (0, 1)$ such that*

$$d(T(x), T(y)) \le \kappa d(x, y)$$

for all $x, y \in X$.

Note that, by the definition, a contraction is necessarily a continuous function on X.

THEOREM A.14. (Banach's Fixed Point Theorem) *If T is a contraction defined on a complete metric space (X, d), then T has a unique fixed point in X.*

Proof. We start by proving the existence of a fixed point. Take an arbitrary point $x_1 \in X$ and consider its successive images under T:

$$x_{n+1} = T(x_n), \qquad n = 1, 2, \ldots.$$

We claim that these elements form a Cauchy sequence. Thus, since X is complete, $x^* = \lim_{n \to \infty} x_n$ exists and belongs to X. The point x^* is a fixed point of T, since for any $y \in X$ and any integer n we have by the triangle inequality

$$d(x^*, T(x^*)) \leq d(x^*, T^n(y)) + d(T^n(y), T^{n+1}(y)) + d(T^{n+1}(y), T(x^*))$$

$$\leq (1 + \kappa)d(x^*, T^n(y)) + \kappa^n d(y, T(y)).$$

Since $d(x^*, T^n(y)) \to 0$ as $n \to \infty$, we have $x^* = T(x^*)$. To verify our claim that $\{x_n\}_{n=1}^{\infty}$ is a Cauchy sequence, it suffices to show that given any $\varepsilon > 0$, there is an $N \in \mathbb{N}$ such that

$$d(x_n, x_{n+p}) < \varepsilon \quad \text{for all } n > N, \quad p = 1, 2, \ldots.$$

Using the triangle inequality again, we get an upper bound on $d(x_n, x_{n+p})$:

$$d(x_n, x_{n+p}) \leq d(x_n, x_{n+1}) + d(x_{n+1}, x_{n+2}) + \cdots + d(x_{n+p-1}, x_{n+p}).$$

Using the contraction hypothesis, we see that

$$d(x_m, x_{m+1}) \leq \kappa d(x_{m-1}, x_m) \leq \cdots \leq \kappa^{m-1} d(x_1, x_2),$$

and therefore

$$d(x_n, x_{n+p}) \leq (\kappa^{n-1} + \kappa^n + \cdots + \kappa^{n+p-1})d(x_1, x_2) < \frac{\kappa^{n-1}}{1 - \kappa}d(x_1, x_2).$$

This expression can be made as small as we wish, simply by taking n sufficiently large. Hence $\{x_n\}_{n=1}^{\infty}$ is a Cauchy sequence as claimed, and its limit x^* exists.

Turning to the question of uniqueness, suppose that z^* also is a fixed point. Then we have

$$d(x^*, z^*) = d(T(x^*), T(z^*)) \leq \kappa d(x^*, z^*).$$

Since $\kappa < 1$, this implies that $d(x^*, z^*) = 0$, which means that $x^* = z^*$. Hence the fixed point is unique. $\qquad \square$

Banach's fixed point theorem is used to establish the existence of solutions to ordinary differential equations (see Section 6.2).

Appendix B

Program Codes

B.1 IEEE CONSTANTS

```
1  /*    File: constants.c
2
3        A simple C program that computes the smallest
4        positive machine representable number, Eta, and
5        the machine epsilon, Eps.
6        Compilation: gcc -Wall -o constants constants.c -lm
7
8  */
9
10 #include <stdio.h>
11
12 /* Calculation of Eta = min { x > 0 }. */
13 double Eta()
14 {
15   double Current = 1.0;
16   double Last    = 1.0;
17
18   while ( Current > 0.0 ) {
19     Last = Current;
20     Current /= 2.0;
21   }
22
23   return Last;
24 }
25
26 /* Calculation of Eps = min { x >= 0 : 1 + x > 1 }. */
27 double Eps()
28 {
29   double Current          = 1.0;
30   double Last             = 1.0;
31   double OnePlusCurrent   = 2.0;
32
33   while ( OnePlusCurrent > 1.0 ) {
34     Last = Current;
35     Current /= 2.0;
36     OnePlusCurrent = 1.0 + Current;
37   }
38
39   return Last;
40 }
41
42 int main()
43 {
```

```
44   printf("eta = %17.17e\n", Eta());
45   printf("eps = %17.17e\n", Eps());
46
47   return 0;
48 }
```

B.2 CHANGING ROUNDING MODES

```
1  /*   File: round.h
2
3        A header file that defines the directed
4        rounding modes for Linux and Sparc.
5  */
6
7  #if defined(__linux)
8  #include <fenv.h>
9  #define ROUND_DOWN   FE_DOWNWARD
10 #define ROUND_UP     FE_UPWARD
11 #define ROUND_NEAR   FE_TONEAREST
12 #define setRound     fesetround
13 #endif
14
15 #if defined(__sparc)
16 #include <ieeefp.h>
17 #define ROUND_DOWN   FP_RM
18 #define ROUND_UP     FP_RP
19 #define ROUND_NEAR   FP_RN
20 #define setRound     fpsetround
21 #endif
22
23 void setRoundDown() { setRound(ROUND_DOWN); }
24 void setRoundUp   () { setRound(ROUND_UP);   }
25 void setRoundNear() { setRound(ROUND_NEAR); }
```

B.2.1 A *mex* file for *MATLAB*

```
1  /*   File: setround.c
2
3        A simple C program that allows directed rounding
4        from within the matlab environment.
5
6        Compilation: mex setround.c
7  */
8
9  #include "mex.h"
10 #include "round.h"
11
12 void round(double *In)
13 {
14     if ( In[0] == mxGetInf() )
15         setRoundUp();
16     else if ( In[0] == -mxGetInf() )
```

```
17            setRoundDown();
18        else if ( In[0] == 0.0 )
19            setRoundZero();
20        else if ( In[0] == 0.5 )
21            setRoundNear();
22        else
23            mexErrMsgTxt("Valid args: +inf, -inf, 0, 0.5.");
24 }
25
26 void mexFunction(int nlhs, mxArray *plhs[],
27                  int nrhs, const mxArray *prhs[])
28 {
29     mxArray *In_ptr;
30     double  *In;
31
32     In_ptr = (mxArray *)prhs[0];
33     In = mxGetPr(In_ptr);
34
35     if ( nrhs != 1 )
36         mexErrMsgTxt("Valid args: +inf, -inf, 0, 0.5.");
37     else if ( nlhs != 0 )
38         mexErrMsgTxt("No output argument allowed!");
39     round(In);
40 }
```

B.3 A SAMPLE CODE IN C++

```
1 /*    File: IAexCC.cc
2
3      A simple C++ program that illustrates interval
4      arithmetic with directed rounding.
5      Compilation: g++ -Wall -o IAexCC IAexCC.cc
6
7 */
8
9 #include <iostream.h>
10 #include <stdlib.h>
11 #include "round.h"
12
13 double Min(double dbl1, double dbl2)
14 { return (dbl1 < dbl2 ? dbl1 : dbl2); }
15
16 double Max(double dbl1, double dbl2)
17 { return (dbl1 > dbl2 ? dbl1 : dbl2); }
18
19 class interval
20 {
21   friend interval operator + (interval iv1, interval iv2);
22   friend interval operator - (interval iv1, interval iv2);
23   friend interval operator * (interval iv1, interval iv2);
24   friend interval operator / (interval iv1, interval iv2);
25   friend ostream & operator << (ostream &, interval);
26   friend istream & operator >> (istream &, interval &);
27 public:
```

```
28    interval() {}
29    interval(double min, double max) {lo = min; hi = max;}
30
31 private:
32    double lo;
33    double hi;
34 };
35
36 interval operator + (interval iv1, interval iv2)
37 {
38    interval result;
39
40    setRoundDown();
41    result.lo = iv1.lo + iv2.lo;
42    setRoundUp();
43    result.hi = iv1.hi + iv2.hi;
44    setRoundNear();
45
46    return result;
47 }
48
49 interval operator - (interval iv1, interval iv2)
50 {
51    interval result;
52
53    setRoundDown();
54    result.lo = iv1.lo - iv2.hi;
55    setRoundUp();
56    result.hi = iv1.hi - iv2.lo;
57    setRoundNear();
58
59    return result;
60 }
61
62 interval operator * (interval iv1, interval iv2)
63 {
64    interval result;
65    double temp1, temp2;
66
67    setRoundDown();
68    temp1 = Min(iv1.lo * iv2.lo, iv1.lo * iv2.hi);
69    temp2 = Min(iv1.hi * iv2.lo, iv1.hi * iv2.hi);
70    result.lo = Min(temp1, temp2);
71    setRoundUp();
72    temp1 = Max(iv1.lo * iv2.lo, iv1.lo * iv2.hi);
73    temp2 = Max(iv1.hi * iv2.lo, iv1.hi * iv2.hi);
74    result.hi = Max(temp1, temp2);
75    setRoundNear();
76
77    return result;
78 }
79
80 interval operator / (interval iv1, interval iv2)
81 {
82    interval inv;
83
84    if ( iv2.lo * iv2.hi <= 0.0 ) {
```

```
85       cerr << "Error: division by zero. Bye!" << endl;
86       exit(1);
87     }
88     setRoundDown();
89     inv.lo = 1.0 / iv2.hi;
90     setRoundUp();
91     inv.hi = 1.0 / iv2.lo;
92     setRoundNear();
93
94     return (iv1 * inv);
95   }
96
97   ostream & operator << (ostream &os, interval iv)
98   {  return os << '[' << iv.lo << ',' << iv.hi << ']'; }
99
100  istream & operator >> (istream &is, interval &iv)
101  {
102     double min, max;
103
104     is >> min;
105     is >> max;
106     iv = interval(min, max);
107     return is;
108  }
109
110  int main()
111  {
112     interval iv1(0.3, 2.0);
113     interval iv2(1.5, 2.87);
114     cout.precision(17);
115     cout.setf(ios::showpos);
116     cout.setf(ios::scientific);
117
118     cout << "iv1 = " << iv1 << endl;
119     cout << "iv2 = " << iv2 << endl;
120     cout << "iv1 + iv2 = " << iv1 + iv2 << endl;
121     cout << "iv1 - iv2 = " << iv1 - iv2 << endl;
122     cout << "iv1 * iv2 = " << iv1 * iv2 << endl;
123     cout << "iv1 / iv2 = " << iv1 / iv2 << endl;
124
125     return 0;
126  }
```

Bibliography

[Ab88]   Aberth, O. Precise Numerical Analysis. Wm. C. Brown Publishers, Dubuque, 1988.

[Ab98]   Aberth, O. Precise Numerical Methods Using C++. Academic Press, New York, 1998.

[AH83]   Alefeld, G., Herzberger, J. Introduction to Interval Computations. Academic Press, New York, 1983.

[Ba22]   Banach, S. *Sur les opérations dans les ensembles abstraits et leur application aux équations intégrales*, Fundamenta Mathematicæ 3, 133–81, 1922.

[Be97]   Berz, M. *From Taylor Series to Taylor Models*, in Beam Stability and Nonlinear Dynamics, AIP Conference Proceedings 405, 1997.

[BS97]   Bendtsen, C., Stauning, O. *TADIFF, A Flexible C++ Package for Automatic Differentiation*, Technical Report, IMM-REP-1997-07, Lyngby, 1997.

[BM98]   Berz, M., Makino, K. *Verified Integration of ODEs and Flows Using Differential Algebraic Methods on High-Order Taylor Models*, Reliable Computing 4, 361–69, 1998.

[Br10]   Brouwer, L.E.J. *Über Abbildung von Mannigfaltigkeiten*, Mathematische Annalen 71, 97–115, 1910.

[Co77]   Corliss, G. *Which Root Does the Bisection Algorithm Find?* SIAM Review 19, 325–27, 1997.

[CV01]   Cuyt, A., Verdonk, B., Kuterna, P. *A Remarkable Example of Catastrophic Cancellation Unraveled*, Computing 66, 309–20, 2001.

[CXSC]   CXSC—C++ eXtension for Scientific Computation, version 2.0. Available at http://www.math.uni-wuppertal.de/org/WRST/xsc/cxsc.html.

[Dw51]   Dwyer, P. S. Linear Computations. John Wiley, New York, 1951.

[EW02]   Eugene, L., Walster, W. G. *Rump's Example Revisited*, Reliable Computing 8, 245–48, 2002.

[FS97]   de Figueiredo, L. H., Stolfi, J. Métodos numéricos auto-validados e aplicações. Braz. Math. Colloq. 21, IMPA, Rio de Janeiro, 1997.

[GM03]   Gabai, D., Meyerhoff, G. R., Thurston, N. *Homotopy Hyperbolic 3-Manifolds Are Hyperbolic*, Annals of Mathematics 157:2, 335–431, 2003.

[Ga96]   Garnatz, P. G. *CRAY T90 Series IEEE Floating Point Migration Issues and Solutions*, Cray User Group 1996 Spring Proceedings, 345–49, 1996.

[Go91]   Goldberg, D. *What Every Computer Scientist Should Know about Floating-Point Arithmetic*, Computing Surveys 23:1, 5–48, 1991.

[Gr00]   Griewank, A. Evaluating Derivatives. SIAM, Philadelphia, 2000.

[HH95]   Hammer, R. et al. C++ Toolbox for Verified Computing. Springer-Verlag, Berlin, 1995.

[Ha65]   Hansen, E. *Interval Arithmetic in Matrix Computations*, J. SIAM Numer. Anal., Ser. B 2:2, 308–20, 1965.

[HW04]   Hansen, E., Wallster, G. W. Global Optimization Using Interval Analysis, 2nd ed. Marcel Dekker, New York, 2004.

[Ha95]   Hass, J., Hutchings, M., Schlafly, R. *The Double Bubble Conjecture*, Electronic Research Announcements of the AMS 1, 98–102, 1995.

[Hi96]   Higham, N. J. Accuracy and Stability of Numerical Algorithm. SIAM, Philadelphia, 1996.

[Hi76]   Hille, E. Ordinary Differential Equations in the Complex Domain. John Wiley, New York, 1976.

[IE85]   IEEE Standard for Binary Floating-Point Arithmetic. ANSI/IEEE Std 754-1985, 1985.

[IE08]   IEEE Standard for Floating-Point Arithmetic. ANSI/IEEE Std 754-2008, 2008.

[IE87]   IEEE Standard for Radix-Independent Floating-Point Arithmetic. ANSI/ IEEE Std 854-1987, 1987.

[JK01]   Jaulin, L., Kieffer, M., Didrit, O. Applied Interval Analysis. Springer-Verlag, London, 2001.

[JZ05]   Jorba, À., Zou, M. *A Software Package for the Numerical Integration of ODE by Means of High-Order Taylor Methods*, Experimental Mathematics 14, 99–117, 2005.

[Ke96]   Kearfott, R. B. Rigorous Global Search: Continuous Problems. Kluwer Academic Publishers, Dordrecht, Netherlands, 1996.

[Kn98]   Knuth, D. E. The Art of Computer Programming. 3rd ed. Addison-Wesley, Reading, MA, 1998.

[Ko02]   Koren, I. Computer Arithmetic Algorithms. 2nd ed. A. K. Peters Ltd., Natick, MA, 2002.

[Kr69]   Krawczyk, R. *Newton-Algorithmen zur Bestimmung von Nullstellen mit Fehlerschranken,* Computing 4, 187–201, 1969.

[KM81]   Kulisch, U. W., Miranker, W. L. Computer Arithmetic in Theory and Practice. Academic Press, New York, 1981.

[LT06]   Lerch, M., Tischler, G., Wolff von Gudenberg, J. *FILIB++, a Fast Interval Library Supporting Containment Computations,* ACM Trans. Math. Software 32:2, 299–324, 2006.

[Mi65]   Milnor, J.W. Topology from the Differentiable Viewpoint. Princeton University Press, Princeton, 1965.

[Mo65]   Moore, R. E. *The Automatic Analysis and Control of Error in Digital Computations Based on the Use of Interval Numbers,* in Error in Digital Computing, vol. 1. John Wiley, New York, 1965.

[Mo59]   Moore, R. E. Automatic Error Analysis in Digital Computation. Technical Report Space Div. Report LMSD84821, Lockheed Missiles and Space Company, 1959.

[Mo66]   Moore, R. E. Interval Analysis. Prentice-Hall, Englewood Cliffs, NJ, 1966.

[Mo79]   Moore, R. E. Methods and Applications of Interval Analysis. SIAM Studies in Applied Mathematics, Philadelphia, 1979.

[MK09]   Moore, R. E., Kearfoot, B. R., Cloud, M. J. Introduction to Interval Analysis. SIAM Studies in Applied Mathematics, Philadelphia, 2009.

[Mu09]   Muller, J-M., et al. Handbook of Floating-Point Arithmetic. Birkhäuser, Boston, 2009.

[NJ01]   Nedialkov, N., Jackson, K., Pryce, J. *An Effective High-Order Interval Method for Validating Existence and Uniqueness of the Solution of an IVP for an ODE,* Reliable Computing 7, 449–65, 2001.

[Ne04]   Neumaier, A. *Complete Search in Continuous Global Optimization and Constraint Satisfaction,* Acta Numerica 13, 271–369, 2004.

[Ne90]   Neumaier, A. Interval Methods for Systems of Equations. Encyclopedia of Mathematics and Its Applications 37, Cambridge University Press, Cambridge, 1990.

[Ne01]   Neumaier, A. Introduction to Numerical Analysis. Cambridge University
          Press, Cambridge, 2001.

[Ov01]   Overton, M. L. Numerical Computing with IEEE Floating Point Arith-
          metic. SIAM, Philadelphia, 2001.

[PP98]   Petkovic, M., Petkovic, L. Complex Interval Arithmetic and Its Applica-
          tions. Wiley-VCH, Berlin, 1998.

[PrBi]   PROFIL/BIAS—Programmer's Runtime Optimized Fast Interval
          Library/Basic Interval Arithmetic Subroutines. Available at http://www.
          ti3.tu-harburg.de/Software/PROFILEnglisch.html.

[PC06]   Pryce, J. D., Corliss, G. F. *Interval Arithmetic with Containment Sets*,
          Computing 78:3, 251–76, 2006.

[Ra96]   Ratz, D. *Inclusion Isotone Extended Interval Arithmetic—A Toolbox
          Update*, FCINE Report No. 5/1996, Universität Karlsruhe, 1996.

[Ru88]   Rump, S. M. *Algorithms for Verified Inclusions: Theory and Practice*,
          in Reliability in Computing: The Role of Interval Methods in Scientific
          Computing. Academic Press, Boston, 1988.

[Ru99]   Rump, S. M. *INTLAB—INTerval LABoratory*, in Tibor Csendes, ed.,
          Developments in Reliable Computing, 77–104. Kluwer Academic
          Publishers, Dordrecht, 1999. Source code available at http://www.ti3.
          tu-harburg.de/~rump/intlab/.

[SK99]   Schwarz, E. M., Krygowski, C. A. *The S/390 G5 Floating-Point Unit*,
          IBM J. Res. Development 43: 5/6, 707–21, 1999.

[St95]   Strichartz, R. S. The Way of Analysis. Jones and Bartlett, Boston, 1995.

[Su58]   Sunaga, T. *Theory of an Interval Algebra and Its Application to Numeri-
          cal Analysis*, RAAG Memoirs 2, 29–46, 1958.

[SUN7]   Sun Microsystems, Inc. Forte Developer 7: C++ Interval Arith-
          metic Programming Reference. Available at http://docs.sun.com/app/
          docs/doc/816-2465.

[Tu02]   Tucker, W. *A Rigorous ODE Solver and Smale's 14th Problem*, Found.
          Comp. Math. 2:1, 53–117, 2002.

[WH03]   Walster, W. G., Hansen, E. Global Optimization Using Interval Anal-
          ysis. Series in Pure and Applied Mathematics, vol. 264, CRC Press,
          2003.

[Wa56]   Warmus, M. *Calculus of Approximations*, Bulletin de l'Academie Polon-
          aise de Sciences 4:5, 253–57, 1956.

[Wi63]   Wilkinson, J. H. Rounding Errors in Algebraic Processes. Prentice-Hall, Englewood Cliffs, NJ, 1963.

[Yo31]   Young, R. C. *The Algebra of Multi-valued Quantities*, Mathematische Annalen 104, 260–90, 1931.

Index

Milton Keynes UK
Ingram Content Group UK Ltd.
UKHW020927220924
448618UK00006B/195

9 780691 247656